MATH Resource Guide

Edited by
Joanne Corker
Rozanne Lanczak Williams

Contributing Writers
Linda Benton
Gwen Botka
Trisha Callella
Marcia Fries
Mary Kurth

Project Directors
Carolea Williams
Rozanne Lanczak Williams

Art Director
Tom Cochrane

Designed by
Moonhee Pak

Illustrated by
Kathleen Dunne

Photographed by
Michael Jarrett

Special, heartfelt thanks to the many teachers and students across the country who have contributed their wonderful ideas and projects for this book.

CTP ©1996, Creative Teaching Press, Inc., Cypress, CA 90630
Reproduction of activities in any manner for use in the classroom and not for commercial sale is permissible.
Reproduction of these materials for an entire school or for a school system is strictly prohibited.

Table of Contents

Introduction 3
Resource Guide Components 4
About the *Learn to Read Math* Series 5
Building a Balanced Literacy Program 6

Activities

Level I

Barney Bear Gets Dressed 9
The Costume Parade 12
I See Patterns 15
I See Shapes 18
Mr. Noisy's Book of Patterns 21
Our Pumpkin 24
Scaredy Cat Runs Away 27
The Skip Count Song 30
We Can Make Graphs 33
What Comes in Threes? 36
What Time Is It? 39
Who Took the Cookies from the Cookie Jar? 42

Level II

A-Counting We Will Go 45
The Bugs Go Marching 48
The Crayola® Counting Book 51
Five Little Monsters 54
Let's Measure It! 57
Little Number Stories: Addition 60
Little Number Stories: Subtraction 63
Lunch with Cat and Dog 66
The Magic Money Box 69
Spiders, Spiders Everywhere! 72
Ten Monsters in a Bed 75
The Time Song 78

INTRODUCTION

The *Learn to Read Resource Guide* provides a wealth of ideas and activities for integrating the *Math* emergent reader series in a balanced literacy program. The *Learn to Read Math* series includes 24 books that have been carefully developed to provide emergent readers with text they can successfully read on their own. The text reflects math concepts commonly taught in primary grades.

Children learn to read by reading and to write by writing. They develop skills, strategies, and fluency in a language-rich environment where they have many varied opportunities to read, write, listen, and speak. The activities and ideas in this guide will provide you with a wide variety of motivational activities and innovative ideas to support the beginning reader's math learning and reading skill development.

The ideas and projects were created by kindergarten, first-, and second-grade teachers and their students all across the country. Their projects are photographed throughout the guide and include:

- hands-on math activities
- real-life math experiences
- interactive math displays
- text innovations
- class books
- individual student books
- story dramatizations
- oral reports
- puppets
- games
- pocket chart activities
- story murals
- graphing
- collaborative projects

Children who think of themselves as readers and writers, and whose every attempt is encouraged and supported, develop the confidence to take on new reading and writing challenges. By integrating literacy learning with math concepts, you will be able to implement a program customized to your students' needs. We hope this guide will help you create a learning environment where children not only learn *how* to read but grow to *love* reading as well.

RESOURCE GUIDE COMPONENTS

For each of the 24 books in the *Learn to Read Math* series, there are three pages of information and extension activities. Each three-page section includes the following special features:

WRITING FRAMES

In this section, several writing frames modeling patterned language in the book are listed. For more information about using writing frames, see page 8.

MATH CONCEPTS

Math concepts related to the text and illustrations in the book are listed. Concepts listed are those most commonly taught in primary grades.

RELATED SKILLS

Opportunities abound for teaching specific skills and reading strategies within the context of the series. Look at this section when planning instruction and addressing children's reading difficulties. For more information about related skills, see pages 6–7.

PHOTOGRAPHS

A picture is worth a thousand words. Bright, colorful photographs of projects created by K–2 students appear throughout this guide and provide extra clarity to written directions.

SYNOPSIS

A short sentence and two-page book spread serve as a reminder of the book's content.

ACTIVITIES

Unique and creative activities to extend math learning, reading, and writing make up this section. Projects include hands-on math activities, interactive math displays, student-made big books, individual student books, murals, pocket chart activities, wall stories, and more.

MATERIALS

Easy-to-find materials are listed for each activity. Items common to all classrooms, such as scissors, crayons, and glue, are listed as art supplies. Collage materials include buttons, confetti, fabric and paper scraps, wiggly eyes, pipe cleaners, stickers, yarn, glitter, sequins, and dried macaroni.

LITERATURE LINKS

This section includes a list of books related to the themes and content of the *Learn to Read Math* series and the activities presented in this guide.

LEARNING A SKILL

One skill from the Related Skills or Math Concepts list is developed for the book. Look at the great ideas in this section for ways to incorporate specific skill instruction.

LINKING SCHOOL TO HOME

These take-home activities provide a non-threatening invitation to parents to become part of the classroom community. They encourage communication between home and school, help children connect home and school learning, and provide lots of opportunities for children to share and reinforce new skills.

4   Learn to Read Resource Guide • Math   Creative Teaching Press

About the *Learn to Read Math Series*

Lots of reading! Lots of math!

The *Learn to Read Math* series is designed as a flexible resource for your early literacy program. The books have been written and carefully developed to provide emergent readers with text they can successfully read on their own. The engaging stories, along with colorful and appealing illustrations, make reading a fun and enjoyable experience.

The *Learn to Read Math* series consists of 24 student-sized books for emergent readers and 24 matching big books. Their content reflects math concepts commonly taught in primary grades. Counting, shapes, measurement, patterning, money, and more are explored in the text and illustrations. The books are written on two levels:

Level I books contain eight pages of easy-to-read text. Usually one or two lines of text appear on each page. There is one language pattern with no more than two changes on each page.

Level II books contain sixteen pages of slightly more difficult text. Several lines of text appear on each page. The language pattern may change throughout the story but remains highly repetitive.

The following special features in the *Learn to Read Math* series give maximum support to the beginning reader:

- Repetitive, predictable story lines provide instant success.
- Engaging stories with satisfying endings promote reading for meaning, rather than just sounding out words.
- Colorful illustrations in a variety of artistic styles closely match the text and provide added support.
- Large print is clear and well-placed.
- Natural language patterns (what the reader is used to hearing) move the reader easily through the text.
- Easy and fun activities on the inside back cover extend language and math learning.

BUILDING A BALANCED LITERACY PROGRAM

Focusing on Skills and Strategies

The development of skills and strategies is an ongoing part of a balanced literacy program and occurs within the context of the reading and writing children are doing in the classroom. Skills can be taught formally when children are experiencing specific difficulties or when you anticipate difficulty with a particular text. Skills are tools learners use to make sense of a story when they read and to communicate effectively when they write. Most importantly, skills become strategies when learners apply them to solve their reading and writing difficulties. Developing strategies should be the focus of all skill instruction.

The components of a balanced literacy program include:

- reading aloud
- shared reading
- guided reading
- independent reading
- writing aloud
- shared writing
- guided writing
- independent writing

Reading

READING ALOUD

Reading aloud is an important part of a balanced literacy program. Read to children several times a day in the classroom, and encourage parents to spend at least fifteen minutes a day reading to their children at home. Reading aloud makes a significant impact on the developing reading skills of young children. It builds comprehension, vocabulary, and listening skills, and exposes children to good literature written on a level higher than their instructional level.

Throughout this guide, Literature Links provide lists of books related to the theme and content of those in the series. Enrich your program by choosing read-aloud titles that extend student learning. Children will gain information and math understanding they can access when working on their own. For example, by reading aloud and singing several versions of *Ten on a Bed*, you acquaint children with the story, supply math background, and introduce important vocabulary. This introduction will help the child independently read the *Learn to Read* version, *Ten Monsters in a Bed*.

GUIDED READING

During guided reading, work with small groups of children who each have a copy of the same book. Children could also read small copies of a big book from a prior shared reading session. A guided reading session is a good time to model and reinforce emergent-level strategies such as one-to-one correspondence, return sweep, locating known and unknown words, letter/sound correspondence (phonics), context clues, and visual searching.

As children develop fluency, give them a book they haven't read before that matches their instructional level. Have each child work through the text while getting your and other readers' support. Children discuss the strategies that help them solve reading problems. This is where the real work of reading occurs. After several successful readings of the book, children can take the book home to read to parents.

SHARED READING

Shared reading is a powerful tool for teaching children what reading is all about. Children at all developmental levels are invited to join in the reading of a big book, poem, chant, or pocket chart story. Print is enlarged on shared reading material in order to encourage participation by the whole group. Modeling and child participation occur simultaneously. The emphasis during these sessions is on the joy and satisfaction of reading.

Big books in the *Learn to Read* series are designed primarily for shared reading with emergent readers. Use the repetition, rhyme, and predictable sentence patterns in the text, along with the strong support from illustrations, to lead beginning readers through successful reading experiences. Children enjoy reading the big books again and again during shared reading, and they become favorite choices during independent reading.

Use previously-read big books for specific skill instruction or to reinforce math concepts. For example, *The Costume Parade* is a great book to teach high-frequency words *the*, *is*, and *a*, and describing words such as *funny*, *scary*, and *fuzzy*. Integrate math learning by highlighting the ordinal numbers and focusing on sequencing skills.

INDEPENDENT READING

Emergent readers need many opportunities to read independently. Create a print-rich, reader-friendly classroom by making the following materials accessible:

- big books from previous shared reading sessions
- little books mastered during guided reading
- student-created books modeled after shared big books
- previously introduced pocket chart sets
- wall stories, story murals, and poetry charts
- trade books with text suitable for emergent readers

Writing

Reading and writing complement each other in a balanced literacy program. They are mutually supportive processes—growing expertise in one area influences the other. Encourage emergent readers to write through writing aloud and shared, guided, and independent writing sessions.

WRITING ALOUD

Write on a chalkboard or chart in front of children and "think aloud" about the text as you write. This provides a powerful model on how to write and exposes children to writing conventions such as spacing, punctuation, and spelling. Many teachers write the morning message "aloud" (a brief description of what's happening in the classroom or other noteworthy events).

SHARED WRITING

During a shared writing session, students write with you—it is a collaborative effort. As you guide the process, children supply ideas and input. Children at all developmental levels are invited to participate. Shared writing is a good time to write original stories, poems, class news, information books, or about shared experiences such as guest speakers or field trips. Use shared writing to create innovations and retellings of books children enjoyed during shared reading. For frames relating to specific book titles, refer to Writing Frames sections in this guide.

GUIDED WRITING

During a guided writing session, the child does the writing while receiving your and other children's support and guidance. This is where the real work of writing occurs. On the emergent level, the guided writing session may be fairly structured. For example, group members could repeat and write the same sentence of a writing frame. You may comment on what the writers are doing correctly and supply missing elements to complete the sentence.

INDEPENDENT WRITING

A language-rich environment is not complete without lots of opportunities for children to write on their own. Encourage writing with journals, reading response logs, dramatic play centers with writing supplies, classroom mailboxes, student writing boxes, and logs in the math center. The simple text and patterned language in *Learn to Read* books provide a secure and inviting framework for children's written responses. After reading the books, some children will spontaneously adopt the language pattern and write their own versions.

RECOMMENDED READING FOR TEACHERS

Bialostok, Steve. *Raising Readers*. Peguis, 1992.

Fisher, Bobbi. *Thinking and Learning Together: Curriculum and Community in a Primary Classroom*. Heinemann, 1995.

Raymond, Allen (publisher). *Teaching K–8: Professional Magazine for Teachers*. Early Years, Inc.

Routman, Regie. *Invitations: Changing as Teachers and Learners*. Heinemann, 1991.

Barney Bear Gets Dressed

Level 1

Barney Bear is having so much fun, he has to keep changing shirts.

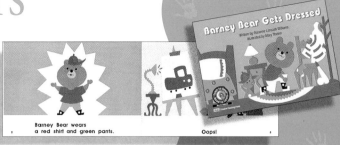

ACTIVITY

WHAT HAPPENED TO BARNEY BEAR?
Story innovation
Cause and effect

A bright idea and project from Tere Fieldson and her kindergartners, Hopkinson School, Los Alamitos, California

Materials
- ✓ blank big book
- ✓ art supplies
- ✓ collage materials

Create a class innovation describing why Barney Bear changed his shirt each time. Use the frame *Barney Bear changed his <u>red</u> shirt because <u>he got paint on it</u>*. Extend the story and vocabulary by having Barney Bear get into new messy situations.

MATH CONCEPTS
- discrete math: *combinations and permutations*
- logical thinking
- problem solving: *creating a strategy*

WRITING FRAMES
____ wears a ____ ____ and ____ ____.

<u>Sam</u> wears a <u>blue</u> <u>shirt</u> and <u>green</u> <u>jeans</u>.

RELATED SKILLS
- cause and effect
- vocabulary: *color words, clothing items*
- punctuation: *exclamation points*

Creative Teaching Press — Learn to Read Resource Guide • *Math*

YUM! ICE CREAM SUNDAES

Story innovation
Food combinations and permutations

Write a class innovation entitled *Barney Bear's Ice Cream Party*. The text can feature combinations of three ice cream flavors and three toppings. On the last page, paste an empty ice cream dish with the text *What other ice cream sundaes can Barney make?* Attach a pocket containing three different ice cream scoops and three more toppings not pictured in the book so children can arrange new combinations.

Materials
- blank class book
- construction paper
- chart paper
- art supplies

Combinations		
Ice Cream	Topping	Initial your choice
chocolate chip	sprinkles	R.W. P.R.
chocolate chip	chocolate syrup	S.P. A.P.
chocolate chip	marshmallow creme	J.R. M.L. J.P.
mint chip	sprinkles	
mint chip	chocolate syrup	
mint chip	marshmallow creme	

Extension Activity

Plan a class ice cream party challenge! Choose three ice cream flavors, three syrups, and three toppings. As a class, invent a strategy for figuring out how many combinations can be made. Work out and chart findings. Have children choose their own combinations by writing their initials next to their choice on a chart. Discuss the most/least favorite combinations and how this affects party planning. Estimate how much of each ice cream flavor, syrup, and topping is needed, enlist parents for help in donating them, and party hearty!

WHAT DO YOU WEAR?

Matching colors with words

A bright idea and project from Marcia Smith and her kindergartners and first graders, Lu Sutton School, Novato, California

Distribute 11" x 17" construction paper and ask each child to draw three outfits for Barney to wear. Each outfit should have a shirt and pants of different colors. Children can write each color word for their clothes on 1" x 4½" construction paper, then glue the words under the correct Barney Bear outfit.

Materials
- 11" x 17" construction paper
- 1" x 4½" construction paper
- art supplies

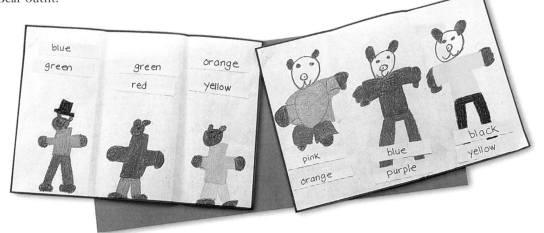

10   Learn to Read Resource Guide • Math   Creative Teaching Press

LEARNING A SKILL

Combinations and permutations

A bright idea and project from Liz Newman and her first graders, Hidden Valley School, Santa Rosa, California

Materials
- construction paper bears
- construction paper bear clothes (blue, red, and yellow shirts; green, purple, and orange pants)
- recording sheets
- art supplies

Provide each pair of children with a bear, a set of clothes, and a recording sheet. Have children manipulate Barney Bear's clothes to investigate how many different outfits they can make. Children can record their findings by coloring bears on the recording sheet, without repeating combinations. Older children can also include blue, yellow, and red hats. Encourage children to share their findings, including the strategy they used when creating different combinations. Store one bear and clothing in a plastic pocket in the back of *Barney Bear Gets Dressed* for children to use when rereading the book.

LINKING SCHOOL TO HOME

Combinations and permutations

Materials
- blank books
- directions sheets

Send home a blank book with each child and a directions sheet. In the directions, ask parents to help children choose three tops and three bottoms and record the various outfits in a blank book entitled *(Child's name) Gets Dressed*. Children can share their books the next day and wear one of the outfits illustrated.

LITERATURE LINKS

Caps for Sale by Esphyr Slobodkina

Caps, Hats, Socks, and Mittens by Louise Borden

Clothesline by Jez Alborough

The Dress I'll Wear to the Party by Shirley Neitzel

Froggy Gets Dressed by Jonathan London

The Jacket I Wear in the Snow by Shirley Neitzel

Jesse Bear, What Will You Wear? by Nancy W. Carlstrom

Martin's Hats by Joan W. Blos

Mary Wore Her Red Dress and Henry Wore His Green Sneakers by Merle Peek

Mouse Paint by Ellen Stoll Walsh

A Three Hat Day by Laura Geringer

You'll Soon Grow Into Them, Titch by Pat Hutchins

Level 1

THE COSTUME PARADE

Children dressed in wonderful costumes help readers explore ordinal numbers.

MATH CONCEPTS

- ordinal numbers: *words and numerals*
- logic

WRITING FRAMES

The _____ costume is a _____.

What a _____ _____!

The <u>first</u> costume is a <u>lion</u>.

What a <u>ferocious</u> <u>lion</u>!

RELATED SKILLS

- high-frequency words: *the, is, a*
- parts of speech: *descriptive words (adjectives)*
- punctuation: *exclamation points*
- sequencing

ON PARADE!
Cartoon strip innovation

A bright idea and project from Candice Siu and her first graders, Lee School, Los Alamitos, California

Materials
- ✓ drawing paper
- ✓ large box with lid
- ✓ two dowels or paper towel tubes
- ✓ butcher paper or white shelf paper
- ✓ art supplies

Create an innovation of *The Costume Parade*, emphasizing ordinal numbers and descriptive words. For example: *The first costume is a cat. What a furry cat!* Have children illustrate one page for each costume and two culminating pages similar to pages seven and eight in the book. Cut a large hole in the face of the box for the "screen." Then cut holes in the top and bottom of the box for the dowels (see photograph). Slide in the dowels and mount the illustrations on a long strip of butcher paper cut to fit the box. Tape the strip to the dowels and turn them to move the story forward.

Learn to Read Resource Guide • *Math* Creative Teaching Press

ACTIVITY: WHERE IS IT?
Paper plate masks
Sequencing

Brainstorm ideas for new characters or costumes, and invite children to make masks from paper plates, construction paper scraps, and collage materials. Tape craft stick "handles" to the masks. Choose five children at a time to stand in a line with their masks. Describe the mask of the first person in line, and have children say the child's name. Invite children to take turns describing a mask and guessing its position in line. For example: *I'm thinking of a green face with large red lips and yellow hair. Where is it? It's the fourth one in line.* Continue the activity until everyone has had a chance to be in the parade.

Materials
- paper plates
- construction paper scraps
- craft sticks
- art supplies
- collage materials

ACTIVITY: TEDDY BEAR PARADE
Class book
Sequencing

A bright idea from Cathy Young, kindergarten teacher, Biella School, Santa Rosa, California

At the beginning of the school year, have children bring teddy bears to school for a parade to meet school personnel. Supply extras if needed. After the parade, have each child draw his or her teddy bear on a bear shape and decorate it with art supplies and collage materials. Award each bear a fun prize that matches its appearance—the fluffiest, the happiest, the most colorful, the softest, and so on. Place bear cutouts in a class book, and help children complete ordinal sentence frames about their teddy bears.

Materials
- construction paper bears
- teddy bear awards
- art supplies
- collage materials

ACTIVITY: THE ORDER OF OUR DAY
Pocket chart activity
Sequencing

Take pictures of your daily routine in the classroom. Write ordinal numbers on word cards, and help children write captions for the photographs on sentence strips. Challenge children to place the photos, word cards, and sentence strips in correct sequence in the pocket chart.

Materials
- camera
- word cards
- sentence strips
- pocket chart

LEARNING A SKILL

Ordinal numbers

Give each child a photocopy of the class picture. Have children choose ten faces to cut out and glue in a line on drawing paper. Children can use markers or crayons to add bodies, clothes, and other details. Each child then completes the sentence frame *One classroom friend is the (ordinal number) person in line. Who is it?* The answer could be added under a flap. Assemble pages in a class book with a laminated construction paper cover composed of children's photos. Children can read the book and guess who is on every page.

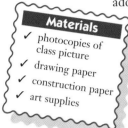

Materials
- ✓ photocopies of class picture
- ✓ drawing paper
- ✓ construction paper
- ✓ art supplies

As an extension, take photographs of children at school for a class book. Print questions beneath each photograph, such as *There are five in line for the bus. Who is third in line?* or *There are three in line for a drink. Who is second in line?*

LINKING SCHOOL TO HOME

Ordinal numbers
Sequencing

As a homework activity, send home a blank accordion book, a copy of the ordinal numbers sheet, and the following note. Allow at least one week for the assignment.

Materials
- ✓ blank accordion books
- ✓ teacher-made ordinal numbers sheets
- ✓ art supplies

Dear Parents,
As an extension of our class lesson on ordinal numbers, please complete the following activity with your child. All family members are invited to participate!

FIRST, choose one activity the family does together, such as planning and making a meal or recipe, writing a family card or letter, or going on an outing.
SECOND, while doing the chosen activity, take notes on what you do first, second, third, and so on.
THIRD, use the blank book and ordinal number sheet to create a sequential story describing your activity. Use photos or drawings to illustrate the book.
FOURTH, have your child bring the finished book to school to share.

Thank you and have fun!

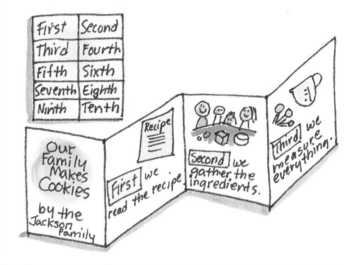

LITERATURE LINKS

The April Rabbits by David Cleveland

Apartment Three by Ezra Jack Keats

Bicycle Race by Donald Crews

Biggest, Strongest, Fastest by Steve Jenkins

The Bus Stop by Nancy Hellen

Freight Train by Donald Crews

Halloween Parade by Harriet Ziefert

The King's Chessboard by David Birch

The Mud Flat Olympics by James Stevenson

Seven Eggs by Meredith Hooper

The Three Billy Goats Gruff (Traditional)

The Three Little Pigs (Traditional)

What Time Is It? by Rozanne Lanczak Williams

I See Patterns

Intriguing photographs help readers identify patterns on objects and in the environment.

Level 1

A_{CTIVITY} LOOK AROUND YOU
Patterns in the environment

Set up a learning center with photography books by Tana Hoban, Bruce McMillan, and other similar books. Children can look for and name patterns in the photos. Also include nature and travel magazines and clothing catalogs in the center. Children can find pictures representing patterns, cut them out, and glue them in a blank book with the frame *I see patterns in the _____*.

Materials
- ✓ *I See Patterns*
- ✓ several books by Tana Hoban and Bruce McMillan (see Literature Links)
- ✓ magazines
- ✓ blank big book
- ✓ art supplies

MATH CONCEPTS
- patterns: *linear, non-linear, symmetrical*
- connecting math to the real world
- shapes
- vocabulary: *horizontal, vertical, diagonal*

WRITING FRAMES

I see patterns on the _____.

I see patterns on the <u>fence</u>.

I hear patterns in the _____.

I hear patterns in the <u>rain</u>.

RELATED SKILLS
- high-frequency words: *I, see, on, the*
- parts of speech: *position words (prepositions)*

A_{CTIVITY} PATTERN MAT PARTNERS
Pattern extensions

Use 9" x 36" construction paper to make pattern mat grids with three rows of twelve 3" squares. Provide pairs of children with a pattern mat and set of manipulatives. Ask one child to start a pattern for his or her partner to finish. Children can also experiment with play mats by adding buttons to shirt shapes and barrettes to head shapes.

Materials
- ✓ 9" x 36" construction paper
- ✓ manipulatives (money, links, pattern blocks, erasers, toy cars, buttons, barrettes)
- ✓ play mats (shirt shapes, head shapes with hair)

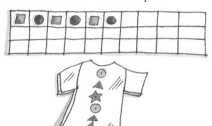

Creative Teaching Press • Learn to Read Resource Guide • *Math*

ACTIVITY: PATTERN PATCHWORK QUILTS
Cooperative paper pattern quilt

A bright idea and project from Carol Pascarella and her kindergartners, St. Marys School, St. Marys, West Virginia

Bring in a real quilt from home and ask children to identify patterns. Also examine the patterns in quilts from books. Distribute 9" paper squares and invite children to create a pattern using multicolored 1" squares of construction paper. Help children arrange their finished squares on a large sheet of butcher paper to form a "quilt." Display the pattern patchwork quilt for children and other classes to enjoy.

Materials
- quilt
- butcher paper
- 9" paper squares
- 1" multicolored construction paper squares
- art supplies

ACTIVITY: EATING UP PATTERNS
Fruit patterns

Ask children to bring in a variety of fresh fruit. Cut fruit into pieces and place them on a large sheet of butcher paper or tablecloth spread out on the floor. With children sitting around the tablecloth, create patterns with the fruit and invite children to name the patterns.

Cut fruit into small pieces, and ask each child to select and thread fruit onto a skewer to create a pattern. Children can draw their fruit patterns on construction paper strips or in their math journals. Invite them to eat their colorful fruit kabobs. Note: Children can also thread fruit pieces on toothpicks to make mini fruit kabobs.

Materials
- fruit
- butcher paper or tablecloth
- knife (teacher use only)
- wooden skewers
- 4" x 18" construction paper strips
- paper plates
- napkins
- toothpicks (optional)
- art supplies

LEARNING A SKILL

Constructing patterns

Materials
- 12" x 18" construction paper
- colored paper geometric shapes
- thematic shapes and stickers (fall leaves, pumpkins)
- art supplies
- shelled peanuts
- vegetables (carrots, celery sticks)
- breakfast cereal
- raisins
- paper plates

A bright idea and project from Caroline Ellis and her kindergartners, Biella School, Santa Rosa, California

Have children use geometric and thematic paper shapes and stickers to create border patterns on 12" x 18" construction paper. Children can share their patterns with friends and ask them to name the patterns.

Children can also create patterns using peanuts, vegetables, breakfast cereal, and raisins. Children can "read" their patterns as they eat their pattern snacks.

LINKING SCHOOL TO HOME

Patterns at home

Give each child a take-home sheet entitled *Patterns in My Home*. Encourage children and parents to look for patterns around the house and record or draw the patterns they find. They can also create patterns with shoes, toys, or kitchen items.

Materials
- 9" x 12" drawing paper
- art supplies

LITERATURE LINKS

The Bedspread by Sylvia Fair

Dots, Spots, Speckles, and Stripes by Tana Hoban

Eating Fractions by Bruce McMillan

Jesse Bear, What Will You Wear? by Nancy White Carlstrom

Look Once, Look Twice by Janet Marshall

The Mountains of Quilt by Nancy Willard

Mouse Views: What the Class Pet Saw by Bruce McMillan

Mr. Nick's Knitting by Margaret Wild

The Patchwork Farmer by Craig Brown

The Patchwork Lady by Mary K. Whittington

Sam Johnson and the Blue Ribbon Quilt by Lisa Campbell Ernst

Spirals, Curves, Fanshapes, and Lines by Tana Hoban

Level 1

I See Shapes

Children's collage art shows shapes of objects at a birthday party.

MATH CONCEPTS

- geometric shapes
- patterns

WRITING FRAMES

I see ____. Now I see ____!

I see <u>ovals</u>. Now I see <u>eggs</u>!

RELATED SKILLS

- high-frequency words: *I, see, now*
- vocabulary: *shape words*
- classifying

TANGRAM TALES
Class book

A bright idea and project from Marcia Fries and her multi-age primary class, Lee School, Los Alamitos, California

Read the books *Grandfather Tang's Story: A Tale Told with Tangrams* and *The Tangram Magician*. While reading, have volunteers form tangrams using felt or magnetic tangram pieces. Place tangram pieces in a center with puzzle solutions so children have many opportunities to make tangrams.

Pair each child with a partner. Have each pair create a page for a class book by making an animal with tangram pieces cut from fluorescent paper and writing a sentence using the frame:

____ and ____ became a ____ so they could ____.

<u>Christiana</u> and <u>Daniel</u> became a <u>fox</u> so they could <u>run fast through the forest</u>.

Add details cut from construction paper scraps to finish the pages.

Materials
- ✓ *Grandfather Tang's Story: A Tale Told with Tangrams* by Ann Tompert
- ✓ *The Tangram Magician* by Lisa C. and Lee Ernst
- ✓ magnetic or felt tangram pieces
- ✓ black butcher paper
- ✓ fluorescent-colored paper
- ✓ tangram patterns
- ✓ construction paper scraps

Learn to Read Resource Guide • Math   Creative Teaching Press

SHAPES ALL AROUND
Shapes mural

A bright idea and project from Gerianne Smith and her kindergartners, Minnie Gant School, Long Beach, California

Materials
- ✓ butcher paper
- ✓ drawing paper
- ✓ art supplies

Invite children to look around the classroom and playground for interesting objects. Brainstorm ideas for shapes within these objects. Have each child paint an object he or she found. Arrange decorated shapes on a large sheet of butcher paper to create a mural. Label each picture and add the following rhymes to corresponding sections of the mural:

What is a triangle?
Can you guess with me?
A triangle has three sides.
Count them—you'll see!

What is a circle?
A circle is round.
Here are some circles
That we just found.

What is a rectangle?
A rectangle has four sides.
Two long and two short.
That is your guide.

What is a square?
A square has four sides, too.
All sides are equal.
That is your clue.

WE SEE SHAPES
Interactive bulletin board

A bright idea and project from Trisha Callella and her first graders, Rossmoor School, Los Alamitos, California

Brainstorm objects to represent each shape in the story. Have each child choose a shape and an object it can become. Help children trace shape patterns on folded construction paper and cut out the shapes, leaving them attached at the fold. On the front, children can paste a sentence strip saying *I see a (shape)*. On the inside, they can paste a sentence strip saying *Now I see a (object)*, and decorate the shape to be their chosen object.

Materials
- ✓ 11" x 17" construction paper
- ✓ shape patterns
- ✓ small sentence strips
- ✓ butcher paper
- ✓ small sponges
- ✓ art supplies

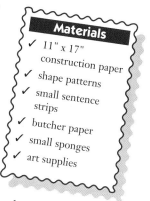

Display work on a bulletin board for children to read and predict what will appear under the flaps. Sponge-paint geometric shapes on butcher paper as a background for the mural.

Creative Teaching Press — Learn to Read Resource Guide • Math

LEARNING A SKILL

Shapes and patterns

A bright idea from Carol Pascarella, St. Marys School, St. Marys, West Virginia

Help each child make a construction paper shape book with the following shape pages: circle, square, triangle, and rectangle. Add a construction paper cover for children to decorate with colored shapes. Invite children to turn shape pages into pictures or paint each page with corresponding sponge shapes. For example, the circle can be a sun or happy face and the square can be a gift or window.

As an extension, enjoy a "shape snack" with children. Supply each child with a baggie containing square, circular, and rectangular crackers and triangular and cubed pieces of cheese. Children can create patterns with crackers and cheese before eating them.

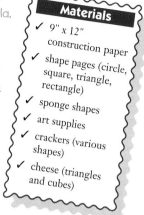

Materials
- ✓ 9" x 12" construction paper
- ✓ shape pages (circle, square, triangle, rectangle)
- ✓ sponge shapes
- ✓ art supplies
- ✓ crackers (various shapes)
- ✓ cheese (triangles and cubes)

LINKING SCHOOL TO HOME

Matching shapes

Place the journal, bag of shape manipulatives, timer, and *I See Shapes* in a decorated tote bag. Have children take turns bringing the bag home. After reading the story together, each family member randomly selects a shape out of the shape bag. They then set the timer and race through the house to collect as many objects of that shape as they can find in the designated amount of time. Have parents help their children count the total number of objects found by family members. Invite them to draw pictures in the class take-home journal of objects found in their house. Family members can sign their family's page and return the journal the next day for another child to take home.

Materials
- ✓ decorated take-home bag
- ✓ journal
- ✓ small bag of shape manipulatives
- ✓ timer
- ✓ *I See Shapes*

LITERATURE LINKS

Changes, Changes by Pat Hutchins

Circles by Arnold Shapiro

Circles, Triangles and Squares by Tana Hoban

Color Farm by Lois Ehlert

Color Zoo by Lois Ehlert

Go Away, Big Green Monster! by Ed Emberley

Grandfather Tang's Story: A Tale Told with Tangrams by Ann Tompert

The Greedy Triangle by Marilyn Burns

Pezzatino by Leo Lionni

Sea Shapes by Suse MacDonald

Shapes in Nature by J. Feldman

Squares by Arnold Shapiro

The Tangram Magician by Lisa C. and Lee Ernst

Triangles by Arnold Shapiro

The Village of Round and Square Houses by Ann Grifalconi

Mr. Noisy's Book of Patterns

Visual and auditory patterns abound in this tale about the lovable Mr. Noisy.

Level I

LOOKING FOR PATTERNS
Identifying and labeling patterns

A bright idea and project from Jennifer Botenhagen and her first graders, Lu Sutton School, Novato, California

Take children on a pattern hunt through *Mr. Noisy's Book of Patterns*, and ask them to look for patterns in the text and pictures. Volunteers can explain patterns such as the *OXOXOX* pattern on Mr. Noisy's tie and recreate them with magnets on a magnet board. Ask children to work alone or with a partner and choose patterns from the book to copy and label. Compile pages into a class book or display patterns on a bulletin board.

Materials
- *Mr. Noisy's Book of Patterns* big book
- magnetic board
- magnetic shapes
- drawing paper
- art supplies

MATH CONCEPTS
- patterns: *describing, extending, analyzing*

WRITING FRAMES

When ____ ____, she/he goes: ____.

When <u>Paul</u> <u>speaks</u>, he goes: <u>Yes, yes, yes</u>!

When <u>Moonbee</u> <u>dances</u>, she goes: <u>Tip! Tap! Tip! Tap</u>!

RELATED SKILLS
- high-frequency words: *Mr., when, he, goes*
- parts of speech: *action words (verbs)*
- sound words (onomatopoeia)
- speech bubbles

NOISY PATTERNS
Recognizing and reproducing patterns

Have children sit in a circle. In the center of the circle place the pattern strips and a variety of musical instruments. Invite children to take turns choosing a pattern strip and "playing" the pattern with one of the instruments. "Audience" members can repeat the pattern by clapping.

Materials
- musical instruments (bells, triangles, tambourines, drums)
- pattern strips from *Learning a Skill* activity

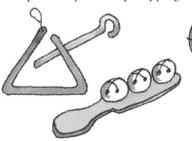

Creative Teaching Press — Learn to Read Resource Guide • Math

ACTIVITY: MR. NOISY'S WORLD OF PATTERNS
Class mural

Draw a simple picture of Mr. Noisy and his dog in the center of a sheet of butcher paper. Assign each child one pattern to add to the mural. This can be done by coloring right on the mural with markers or crayons or by gluing patterns cut from construction paper. Children can add patterns to Mr. Noisy's car, jacket, bow tie, or socks, and create them in trees, flowers, clouds, or in the sidewalk. When the mural is complete, invite children to name the patterns on sentence strips, stick Velcro to the strips and mural, and take turns attaching labels to the patterns.

Materials
- ✓ butcher paper
- ✓ construction paper
- ✓ sentence strips
- ✓ self-adhesive Velcro pieces
- ✓ art supplies

ACTIVITY: MR. NOISY'S NOISY DAY
Action words (verbs)
Onomatopoeia

A bright idea and project from Bev Maeda and her multi-age primary class, Rolling Hills School, Fullerton, California

Explain to students that a verb is an action word. To introduce verbs to beginning readers, use *Add It, Dip It, Fix It: A Book of Verbs*. Twenty-six action words are explained through striking illustrations. Review the story and list some of Mr. Noisy's actions such as *he talks, walks, sings, dances, drives his car, rides his bike,* and *sleeps*. Then brainstorm a list of verbs for Mr. Noisy as he goes through another busy day. Make a list of 20 or more verbs, and have pairs of children choose one action word to act out, illustrate, and place in a sentence, including the appropriate sound words. Print each pair's "Mr. Noisy sentence" on a sheet of paper, using the language pattern from the story. After children draw a picture to match their sentence, combine pages into a class book entitled *Mr. Noisy's Noisy Day*.

Materials
- ✓ *Add It, Dip It, Fix It: A Book of Verbs* by R.M. Schneider
- ✓ *Mr. Noisy's Book of Patterns* big book
- ✓ drawing paper
- ✓ art supplies

LEARNING A SKILL

Exploring and constructing patterns

Review different noisy patterns on each page of the story. Print each pattern on the chalkboard using an ABC labeling code as children describe and analyze the sequences. For example: *Hi Hello Hi Hello* is an AB pattern and *Tip tap tap tap* is an ABBB pattern. As children make noises, print each pattern code, and have children use manipulatives to form patterns on the floor. Have each child choose his or her favorite noisy pattern from the story and print pattern words along the bottom of a construction paper strip. Children can glue manipulatives to the construction paper to represent the pattern. Hang finished strips in the classroom for an eye-catching, hands-on display.

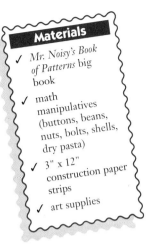

Materials
- Mr. Noisy's Book of Patterns big book
- math manipulatives (buttons, beans, nuts, bolts, shells, dry pasta)
- 3" x 12" construction paper strips
- art supplies

LINKING SCHOOL TO HOME

Patterns at home

Make a book for each child by laying two sheets of drawing paper on top of construction paper. Fold all three pieces in half and staple the folded edge. Cut off the top corners to form a house shape and title each book *Patterns at Home*. Ask parents to help their children look for patterns at home and in their neighborhood. They should try to find at least one "noisy" pattern. Invite children to draw and write about these patterns in their books to share with classmates.

Materials
- drawing paper
- construction paper
- art supplies

LITERATURE LINKS

Add It, Dip It, Fix It: A Book of Verbs by R.M. Schneider

Bam Bam Bam by Eve Merriam

The Banging Book by Bill Grossman

Chicka Chicka Boom Boom by Bill Martin, Jr. and John Archambault

The Color Box by Dayle Ann Dodds

Color Farm by Lois Ehlert

Color Zoo by Lois Ehlert

The Ear Book by Al Perkins

Look Once, Look Twice by Janet Marshall

The Monster Book of ABC Sounds by Alan Snow

Mr. Mumble by Peter Catalanotto

Shhhh by Kevin Henkes

Level 1

OUR PUMPKIN

Real-life photographs portray hands-on pumpkin fun.

MATH CONCEPTS

- estimating
- counting
- measuring: *weight*
- place value
- standard/nonstandard units of measurement

WRITING FRAMES

We can ____ our pumpkin.

We can <u>carve</u> our pumpkin.

We can ____ our ____.

We can <u>graph</u> our <u>apples</u>.

RELATED SKILLS

- high-frequency words: *we, can, our*
- parts of speech: *action words (verbs)*
- phonics: *initial consonant p*
- possessive: *pumpkin's*

PUMPKINS AND APPLES
Weight comparisons

Discuss which is heavier—the apples or pumpkin. Estimate how many apples it would take to weigh the same as the pumpkin, and record everyone's estimations. Place the pumpkin on one side of the balance scale or a bathroom scale. Place the apples on the other side of the balance scale or on the second bathroom scale. Remove or add apples to match their weight with the pumpkin's. Children can count along until the scales are approximately equal. Compare results with estimations and discuss which answers were too much, too little, or close to the actual weight. Leave the scales, pumpkin, apples, and assorted math manipulatives out for children to weigh and compare during free time. They can also weigh themselves and estimate the number of pumpkins or apples it would take to equal their own weight.

Materials
- medium-sized pumpkin
- basket of apples
- balance scale or two bathroom scales
- two tubs of water
- math manipulatives

For outdoor exploration, place the pumpkin in a full tub of water and let the water overflow. Remove the pumpkin and measure the water level. Fill the tub again with the same amount of water. Ask children, *How many apples will we need to add to make the same amount of water spill over again?* Write estimations, perform the activity, and discuss results. Leave supplies outside for free experimentation during recess.

Learn to Read Resource Guide • *Math*

OUR CLASS PUMPKIN
Story innovation

Use the class pumpkin for all activities shown in *Our Pumpkin*. Take photographs as children weigh, measure, float, and cut the pumpkin, as well as when they count, plant, cook, and eat the seeds. Record all information as activities are performed. Make a class book using the photographs. Extend the text by adding information from your recording sheets.

Materials
- pumpkin
- camera
- measuring tools
- bathroom or balance scale
- tub of water
- pumpkin cutter

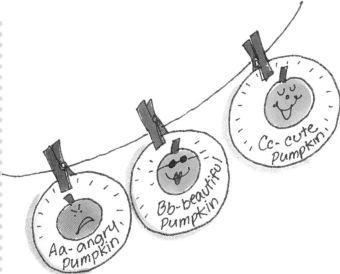

PUMPKINS A TO Z
Descriptive words
Alphabetical order

For each letter of the alphabet, brainstorm silly or serious words to describe a pumpkin. Print letters and describing words on paper plates for children to illustrate. Punch a hole in each plate and string the plates across a bulletin board for children to read.

Materials
- paper plates
- hole punch
- string
- art supplies

GUESS WHAT'S GROWING
Pocket chart activity

A bright idea from Jean Breeze, first-grade teacher, Westview School, Champaign, Illinois

Read *Pumpkin Pumpkin*, sing *The Seed Song* (a *Learn to Read Science* book), or visit a pumpkin patch to look for the growth stages of a pumpkin. Plant pumpkin seeds so children can gain first-hand knowledge of the growth cycle. Write the poem *What Is Growing? Can You Guess?* on sentence strips for the pocket chart. Have children make simple picture cards about pumpkin growth to add to the pocket chart.

Materials
- *Pumpkin Pumpkin* by Jeanne Titherington
- *The Seed Song* by Judy Saksie
- construction paper
- pumpkin seeds
- sentence strips
- pocket chart
- picture cards
- art supplies

What Is Growing? Can You Guess?
by Jean Breeze

Plant a seed. Watch it grow.
What will it be? Do you know?

First, it's a flower. Then a small ball.
It will grow bigger, but not very tall.

At first, it's green. Then to orange it turns.
You cut, clean, and carve. Inside a candle will burn.

Have you guessed? What will be seen?
You're right! A jack-o'-lantern for Halloween.

LEARNING A SKILL

Estimating
Counting

Challenge children to compare the number of seeds in a pumpkin to those in a watermelon, cantaloupe, and/or honeydew. Have children estimate the numbers and suggest which might have the most seeds. They can count the seeds in groups of ten and graph their results. As an extension, children can plant seeds from the pumpkin and melons and chart their growth.

Materials
- pumpkin
- melons (watermelon, cantaloupe, honeydew)
- knife (teacher use only)
- chart paper
- small paper cups
- art supplies

LINKING SCHOOL TO HOME

Number sentences

Provide children with take-home pumpkin-shaped booklets. Ask parents to help their children create number stories using pumpkin, sunflower, or melon seeds. Invite children to paste seeds in their booklets to illustrate addition or subtraction stories. They can attach little paper flaps to lift and reveal answers to their number sentences. Ask children to bring their booklets to school to share with classmates.

Materials
- blank pumpkin-shaped booklets
- pumpkin, sunflower, or melon seeds
- art supplies

LITERATURE LINKS

The Biggest Pumpkin Ever by Steven Kroll

Five Little Pumpkins by Iris Van Rynbach

Grandma's Smile by Elaine Moore

Jack's Garden by Henry Cole

The Pumpkin Blanket by Deborah Turney Zagwÿn

The Pumpkin Patch by Elizabeth King

Pumpkin Pumpkin by Jeanne Titherington

Pumpkins: A Story for a Field by Mary Lyn Ray

Who Sank the Boat? by Pamela Allen

Scaredy Cat Runs Away

Level I

There is a surprise ending when a bear pursues Scaredy Cat across the countryside.

ACTIVITY: THE STORY COMES ALIVE

Story dramatization
Story extensions

A bright idea and project from Marcia Fries and her kindergartners, Lee School, Los Alamitos, California

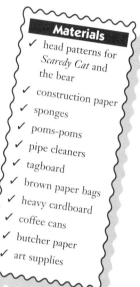

Materials
- head patterns for *Scaredy Cat* and the bear
- construction paper
- sponges
- poms-poms
- pipe cleaners
- tagboard
- brown paper bags
- heavy cardboard
- coffee cans
- butcher paper
- art supplies

Have each child trace and cut out a Scaredy Cat or bear head from white construction paper. Children can sponge-paint fur on their "heads." When dry, they can add wiggly eyes or cutout construction paper eyes. Invite children to glue on large pom-poms for noses, pipe cleaners for whiskers, and add facial details with black markers. Attach each face to the center of a construction paper strip, measure around the child's head, and staple the headband to fit.

To create scenery, cut trees from large pieces of tagboard. Sponge-paint green tops and add crumbled brown paper bags for "tree bark." Use heavy cardboard to make stands for the trees. (Fold cardboard in half and cut slits in the bottom to hold the base of the tree.) Cut out a tagboard fence for children to sponge-paint and add black marker details. Attach two or three strips of heavy cardboard to the back for support. Place fence posts in coffee cans filled with sand. Cut out a butcher paper puddle and have children finger-paint with brown tempera to create "mud."

Place the scenery in position and choose two children to act out the story and one child to narrate. As the child reads the story, the characters can demonstrate directional words. At the end, encourage characters to come alive and say their lines. Children may also wish to write more lines for Scaredy Cat's route, use more directional words, and brainstorm how to make extra scenery.

MATH CONCEPTS

- sequencing
- discrete math
- logical thinking
- measuring: *distance*
- perimeter and diameter

WRITING FRAMES

Scaredy Cat ran ____ the ____.

The bear ran ____ the ____.

Scaredy Cat ran <u>over the bridge</u>.

The bear ran <u>under</u> the <u>bridge</u>.

RELATED SKILLS

- parts of speech: *prepositional phrases*
- punctuation: *proper names (Scaredy Cat vs. bear), quotation marks, exclamation points*
- vocabulary: *directional words, opposites*

Creative Teaching Press   Learn to Read Resource Guide • *Math*

ACTIVITY: SCAREDY CAT KEEPS RUNNING
Story map
Prepositional phrases

A bright idea and project from Renee Keeler and her second graders, Lee School, Los Alamitos, California

Create a bulletin board map of Scaredy Cat's journey, showing the paths of each animal with sponge-stamp footprints. Add prepositional words and phrases such as *over*, *under*, *behind*, and *in front of*. Children can make stick puppets to use for retelling. Brainstorm ideas for new text to go with the story, such as *Scaredy Cat ran over the bridge. The bear ran under the bridge.* Children can also help write a class innovation in which Scaredy Cat chases the bear. Use monthly themes to write a class innovation such as a ghost chasing a witch or a pilgrim chasing a turkey. Print sentences on word cards for children to sequence in the pocket chart during free time.

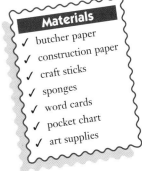

Materials
- butcher paper
- construction paper
- craft sticks
- sponges
- word cards
- pocket chart
- art supplies

ACTIVITY: HOW FAR DID HE GO?
Story map
Length measurements

In small groups, have children draw a story map of *Scaredy Cat Runs Away*. They should include places along Scaredy Cat's route such as the fence, tree, and mud. Children can use string to measure routes and then compare results. They can predict which path is the shortest and check their results with string. Ask children to compare and contrast the ways each group drew their map and compare the lengths of the paths taken. Challenge children to include a legend for their map.

Materials
- *Scaredy Cat Runs Away*
- 12" x 18" butcher paper or construction paper
- string
- art supplies

Learn to Read Resource Guide • Math

LEARNING A SKILL

Sequencing

Cover a shoe box with colored self-adhesive paper. Line the inside lid with felt. Make story scenery and characters from tagboard, construction paper, or felt. If using construction paper or tagboard, glue a piece of felt or sandpaper to the back of each piece so it will stick to the felt background. Have children retell the story, placing corresponding play pieces on the felt lid. Each child can retell one part of the story, then pass the lid to the next person. Store story pieces inside the box for future retellings. Send the story box home with each child so he or she can retell the story to family members.

Materials
- ✓ shoe box
- ✓ colored self-adhesive paper
- ✓ felt
- ✓ construction paper or tagboard
- ✓ art supplies

LINKING SCHOOL TO HOME

Measuring distances

Have each child think of two ways to get from the front of his or her house to the back. Invite children to predict which way is the shortest. They can check their predictions by using their feet to measure each distance. With help from their families, have them draw a map showing both routes. Ask children to bring their maps to school to share with classmates. Have the class guess the shortest routes before children share answers.

Materials
- ✓ drawing paper
- ✓ art supplies

LITERATURE LINKS

Across the Stream by Mirra Ginsburg

All About Where by Tana Hoban

The Bear Went Over the Mountain by Rozanne Lanczak Williams

Bicycle Race by Donald Crews

The Line-up Book by Marisabina Russo

Over in the Meadow by John Langstaff

Over, Under and Through by Tana Hoban

Rosie's Walk by Pat Hutchins

We're Going on a Bear Hunt by Michael Rosen

Level 1

THE SKIP COUNT SONG

A catchy tune and bright photographs help readers count by twos, fives, and tens.

MATH CONCEPTS

- counting by twos, fives, and tens
- number concepts: *combining sets (addition)*
- recognizing numerals
- one-to-one correspondence
- sorting and classifying: *sets and subsets*

WRITING FRAMES

Skip count, skip count, count by ____.

We can count to ____.

Skip count, skip count, count by <u>fours</u>.

We can count to <u>forty</u>.

RELATED SKILLS

- high-frequency words: *we, by, to*
- phonics: *consonant blend sk, vowel digraph ou*
- vocabulary: *number words*

 SKIP COUNT COLLAGE QUILTS
Counting by twos, fives, and tens
Class number quilt

A bright idea from Mary Kurth and Bob Walker and his second graders, Black Earth School, Black Earth, Wisconsin

Cut tagboard sheets in half and prepare ten note cards per quilt section. On two sets of cards, print the numerals *2, 4, 6, . . . 20*. On two other sets of cards, print *5, 10, 15, . . . 50*. On two more sets of cards, print *10, 20, 30, . . . 100*. Divide the class into six groups and provide each group with a different quilt item, a set of ten note cards, and a sheet of tagboard.

Have children adhere two, five, or ten stickers (or other materials) to each note card, then glue the cards in sequence to their tagboard sheet (see photograph). Have children draw a colorful border around each note card using crayons or markers. Once all quilt sections are complete, attach them with masking tape on the back or cover them with clear self-adhesive paper. Hang this colorful counting quilt in the classroom for children to use when practicing skip counting by twos, fives, or tens. Be sure children sign their names on their sections of the quilt.

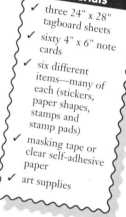

Materials

- ✓ three 24" x 28" tagboard sheets
- ✓ sixty 4" x 6" note cards
- ✓ six different items—many of each (stickers, paper shapes, stamps and stamp pads)
- ✓ masking tape or clear self-adhesive paper
- ✓ art supplies

Learn to Read Resource Guide • Math — Creative Teaching Press

OUR CLASS COUNTS!
Class touch-and-feel book

A bright idea and project from Carol Murakoshi and her kindergartners, Los Alamitos Elementary, Los Alamitos, California

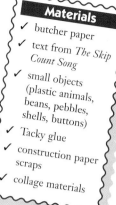

Materials
- ✓ butcher paper
- ✓ text from *The Skip Count Song*
- ✓ small objects (plastic animals, beans, pebbles, shells, buttons)
- ✓ Tacky glue
- ✓ construction paper scraps
- ✓ collage materials

Use butcher paper to make a big book version of *The Skip Count Song*. Attach text to the left side of each page spread. On the right side of each spread, children can glue small objects to match the counting pattern in the text. Invite children to decorate each page with a patterned border made from construction paper scraps and collage materials.

Tape-record children singing and skip counting, and place the big book and cassette together in the listening center.

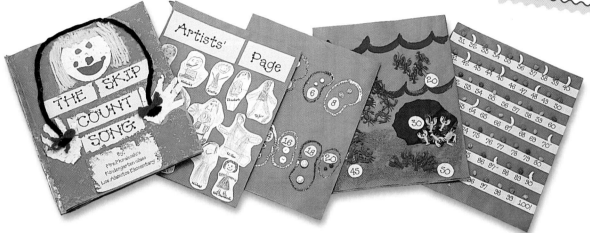

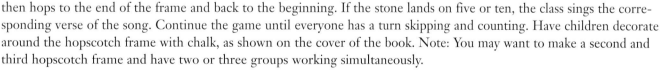

HOPSCOTCH "TOSS AND SKIP COUNT"
Counting by twos, fives, and tens

Materials
- ✓ colored chalk
- ✓ pebble or stone

Take children outside to a sidewalk or black-topped area. Draw a simple hopscotch frame and in each box, print the numeral *2*, *5*, or *10*. Invite children to line up and take turns tossing the stone onto the hopscotch frame. If the stone lands on a two, the class sings *The Skip Count Song* for the number two, and the child hops to the space and picks up the stone. When the class stops singing, the child counts by twos to 20. He or she then hops to the end of the frame and back to the beginning. If the stone lands on five or ten, the class sings the corresponding verse of the song. Continue the game until everyone has a turn skipping and counting. Have children decorate around the hopscotch frame with chalk, as shown on the cover of the book. Note: You may want to make a second and third hopscotch frame and have two or three groups working simultaneously.

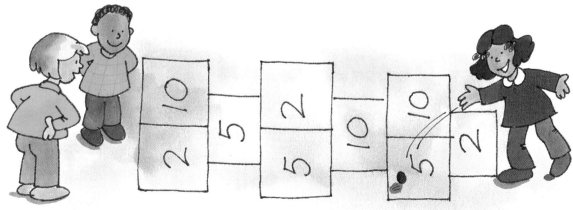

LEARNING A SKILL

Counting by twos, fives, and tens

Have children sit in a circle on the floor. Display the hundreds chart and count aloud together by twos, fives, and tens, as modeled in the book. Circle the numbers on the chart, using a different color marker each time you count. Hand the stone to one child and have him or her begin counting by twos. Have children pass the stone around the circle, each child saying one number until 20 is reached, then beginning again with two. Ring the bell at any stage and ask the child with the stone to go to the center of the circle and count out 20 manipulatives by twos. Repeat the activity a number of times, then do it again, counting with fives and tens.

Materials
- hundreds chart
- stone
- bell
- manipulatives
- art supplies

LINKING SCHOOL TO HOME

Skip count accordion books

Accordion-fold each paper strip to form ten sections of equal size. Fold up the booklet and print *Skip Counting* on the cover. Use a paper clip to shut the book. Have each child take home his or her book and work with a parent to decide whether to count by twos, fives, or tens. Have children draw objects and print numerals in each section of their accordion books, skip counting by the chosen number. At school, the class can skip count in unison as each child points to the pages in his or her book.

Materials
- 4" x 36" construction paper strips
- paper clips
- art supplies

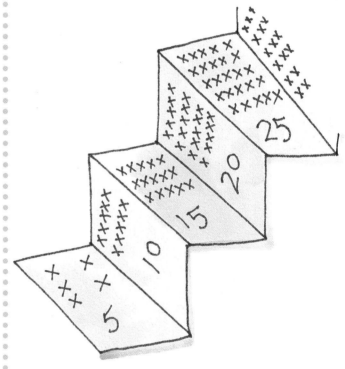

LITERATURE LINKS

Across the Stream by Mirra Ginsberg

Carousel by Donald Crews

Count! by Denise Fleming

Count and See by Tana Hoban

Demi's Count the Animals 1, 2, 3 by Demi

How Many Feet in the Bed? by Dianne Johnston Hamm

How Many Snails? A Counting Book by Paul Giganti, Jr.

Ocean Parade: A Counting Book by Patricia MacCarthy

One Ballerina Two by Vivian French

One Hungry Monster: A Counting Book in Rhyme by Susan Heyboer O'Keefe

The 329th Friend by Marjorie Weinman Sharmat

Two by Two: The Untold Story by Kathryn Hewitt

Two Ways to Count to Ten by Ruby Dee

Up to Ten and Down Again by Lisa Campbell Ernst

What Comes in 2s, 3s, and 4s? by Suzanne Aker

WE CAN MAKE GRAPHS

Children make a different graph every day of the week.

Level 1

ACTIVITY: WHAT'S YOUR FAVORITE SPORT?
Recording and interpreting graph data

A bright idea and project from Lori Avalos and her second graders, Los Alamitos Elementary, Los Alamitos, California

Materials
- ✓ butcher paper
- ✓ blank paper doll shapes
- ✓ art supplies

Discuss which sports children play and enjoy the most. Choose four or five sports to graph on a butcher paper mural. Children can pick their favorite sports and decorate paper doll shapes to represent themselves playing those sports. Have children glue their paper dolls to the appropriate section of the graph. Discuss the results with questions such as *Which sport is most/least popular? How many more girls than boys like a particular sport?* Add this information to the graph in star shapes.

ACTIVITY: HOW DO WE GET TO SCHOOL?
Recording and interpreting graph data

A bright idea from Karri Haven, kindergarten teacher, Helen Lehman School, Santa Rosa, California

Materials
- ✓ small cutouts (buses, cars, feet, bikes)
- ✓ chart paper
- ✓ art supplies

Invite children to choose a cutout showing how they get to school. They can write their names on their cutouts and place them in the matching column on a large graph. Ask children to count and compare their results.

MATH CONCEPTS
- counting
- graphing: *organizing and interpreting data*
- comparing: *more, fewer, equal*
- measuring time: *days of the week*
- one-to-one correspondence
- sequencing

WRITING FRAMES

On _____ we made a graph about _____.

On <u>Saturday</u> we made a graph about <u>our favorite stories</u>.

We can graph _____.

We can graph <u>seeds</u>.

RELATED SKILLS
- days of the week
- high-frequency words: *we, can, make, a*
- parts of speech: *prepositions*
- phonics: *consonant digraph ph*

Creative Teaching Press — Learn to Read Resource Guide • Math

ACTIVITY: HOW DO YOU DECORATE A GINGERBREAD PERSON?
Recording and interpreting data
Cooking

A bright idea from Linda Palmer, kindergarten teacher, Mulberry School, Whittier, California

After reading the recipe together, have children help with each step of cookie preparation. Invite them to choose toppings to decorate their cookies. Ask children to glue the same toppings to their paper gingerbread people (or represent decorations using crayons or markers). When finished, help children place their paper gingerbread people on a graph. Discuss favorite ways to decorate cookies, using terms such as *more, less, most, least, how many more,* and *how many less.*

Materials
- ✓ box of gingerbread cookie mix or favorite recipe
- ✓ baking supplies
- ✓ gingerbread people cookie cutters
- ✓ decorating materials (small candies, raisins, frosting)
- ✓ blank construction paper gingerbread people
- ✓ chart paper
- ✓ art supplies

ACTIVITY: WHAT'S YOUR FAVORITE?
Graphing choices
Graph interpretation

A bright idea and project from Candice Siu and her first graders, Lee School, Los Alamitos, California

Cut different types of apples into small slices and arrange them on paper plates for children to taste test. Ask them to choose their favorite apple, take an apple cutout to match their selection, and glue it onto the graph. Ask, *What can you tell from the graph?* Write this information in bubbles cut from construction paper and glue them onto the graph for children to reread during free reading.

Materials
- ✓ variety of apples
- ✓ knife (teacher use only)
- ✓ paper plates
- ✓ apple cutouts (green, yellow, red)
- ✓ large graph with color labels
- ✓ construction paper
- ✓ art supplies

LEARNING A SKILL

Counting and graphing results

Gather items that can be sorted by categories such as size, color, and shape. Fill each jar with different amounts of objects from 30 through 200. Give a counting jar to each child or small group of children and ask them to decide how to sort their objects, such as by size or shape. Children can make a graph to illustrate sorting techniques and share results with the class.

Materials
- small jars
- counters (dry beans, plastic spiders, shells, candies, dry pasta)
- chart paper

LINKING SCHOOL TO HOME

Estimating
Counting

Help each child accordion-fold 9" x 12" construction paper to make a four-page take-home book entitled *Fruit Market Math*. Ask parents and children to examine several pieces of fruit. Children can estimate which weighs the most and weigh each piece to find the heaviest. They can estimate which fruit has the most seeds, and parents can help cut the fruit for them to count the seeds. Children can then taste the fruit and choose their favorite. Have parents help children complete the following sentence frames on each page of their *Fruit Market Math* booklets:

The _____ weighed the most.
The _____ had the most seeds.
The _____ was my favorite.

After children return their booklets to school, use the information to complete a class graph about the fruit used. Children can use fruit stickers or cutouts to record results.

Materials
- 9" x 12" construction paper
- fresh fruit (apples, pears, oranges)
- knife (adult use only)
- kitchen or balance scale
- chart paper
- art supplies

LITERATURE LINKS

Caps for Sale by Esphyr Slobodkina

Dad's Diet by Barbara Comber

Hamet's Halloween Candy by Nancy Carlson

Martin's Hats by Jan Bios

A More or Less Fish Story by Joanne and David Wylie

Mr. Archimedes' Bath by Pamela Allen

1 Hunter by Pat Hutchins

Red Is Best by Kathy Stinson

When the King Rides By by Margaret Mahy

When We Went to the Park by Shirley Hughes

Level 1

WHAT COMES IN THREES?

Famous storybook characters are revisited in this fun-to-read book about threes.

MATH CONCEPTS
- number sense and numeration
- classifying
- counting

WRITING FRAMES

There are three ____.

There are two ____.

But there's only one ____!

____ come in ____.

Shoes come in *twos*.

RELATED SKILLS
- high-frequency words: *there, are*
- oral language: *retelling fairy tales*

HANG TWO OR MORE
Number mobiles

A bright idea and project from Linda Goodner and her kindergartners, Lee School, Los Alamitos, California

Divide children into small groups and give each group a tagboard number. Have groups decorate one side of their number with collage materials such as confetti, macaroni, sequins, or stickers. As numbers dry, children can search for magazine pictures that go with their numbers, such as two eyes for the number *two* or a four-legged dog for *four*. Help children glue pictures to precut circles and label them. Punch holes in the bottoms of the tagboard numbers and attach the circles with string. Have children share their findings with the class and display their projects around the room.

Materials
- ✓ large tagboard numbers
- ✓ magazines
- ✓ 5" colored circles
- ✓ hole punch
- ✓ string
- ✓ art supplies
- ✓ collage materials

Learn to Read Resource Guide • *Math*

ACTIVITY

MY NUMBER BOOK
Individual paper bag books

A bright idea and project from Marcia Fries and her kindergartners, Lee School, Los Alamitos, California

Give each child four paper bags and attach them on the left side with staples or punch holes and tie yarn to make a binding. On the cover, have children write the title *My Number Book*. They can decorate the cover with small cutout or sponge-painted numbers. On the next page, have children write *What comes in ____?* Let them choose a number to put in the blank. Then give children magazines and have them search for pictures to put inside their bags. When they have completed all three subsequent bags, invite them to mix the items and place them in the first bag (the cover). Challenge children to sort their pictures and place them in the correct bags. Invite them to share their bags with a friend.

Materials
- small paper bags
- binding materials (yarn, staples)
- hole punch
- small tagboard number patterns
- small sponges
- magazines
- art supplies

ACTIVITY

WHAT COMES IN TWOS?
Classifying

A bright idea and project from Kimberly Jordano and her kindergartners, Lee School, Los Alamitos, California

Share *What Comes in Threes?* and *What Comes in 2s, 3s, and 4s?* with the class. Brainstorm things that come in twos. Help children accordion-fold sheets of construction paper into six sections and provide each child with a copy of the following poem:

Hands and feet come in 2s.
Eyes and ears—they do, too!
Lips come in 2s, to make you smile,
But there's only one (child's name) to love!

Help children paste a sentence on each page of their book, then add appropriate illustrations such as handprints, construction paper feet, eyes, ears, and a lip-stamp smile. Place a photo of each child on the last page of his or her book. Finished books make great gifts for parents.

Materials
- *What Comes in Threes?*
- *What Comes in 2s, 3s, and 4s?* by Suzanne Aker
- construction paper
- individual student photos
- art supplies

ACTIVITY

CAN YOU REMEMBER?
Memory game

Brainstorm ideas for a memory game using three objects that can be grouped together. Provide each child with eighteen tagboard pieces. Have children draw one each of the following "threes" story characters: bears, billy goats, pigs, kittens, pairs of mittens, and magic beans. Paste each picture on a 3" x 4" game card. Children can then play a memory game with a friend or by themselves, finding sets of three with their cards. Have children store game cards in plastic zipper bags.

Materials
- 3" x 4" tagboard pieces
- drawing paper
- plastic zipper bags
- art supplies

Creative Teaching Press • Learn to Read Resource Guide • Math

LEARNING A SKILL

Matching numerals and numbers of objects

Help each child trace and cut out a number on construction paper. Have children glue the matching number of objects to their numbers, leaving spaces for cutting between various items. When numbers are dry, have children cut the numbers apart in various zig zags to make puzzle pieces. Challenge children to put their number puzzles back together. Store puzzle pieces in envelopes and invite children to trade with classmates.

Materials
- ✓ large tagboard number patterns
- ✓ construction paper
- ✓ small objects (magazine pictures, stickers, buttons)
- ✓ envelopes
- ✓ art supplies

LINKING SCHOOL TO HOME

Flip-flap books

Make a flip-flap book for each child as follows:

1. Fold 9" x 12" construction paper in half lengthwise.

2. At the top, write *What Comes in ____?* Fold the top flap underneath the words.

3. Cut four equal sections from the bottom to fold under the words on the top flap only.

Send home a flip-flap book with each child. Have children find something in their houses to go under each number. Ask children to draw a picture for each number and bring their books back to school to share with classmates.

Materials
- ✓ 9" x 12" construction paper
- ✓ art supplies

LITERATURE LINKS

Across the Stream by Mirra Ginsberg

Carousel by Donald Crews

Count and See by Tana Hoban

Demi's Count the Animals 1, 2, 3 by Demi

Only One by Marc Harshman

Scarry's Best Counting Book Ever by Richard Scarry

The Snow Parade by Barbara Brenner

The Three Billy Goats Gruff (Traditional)

The 329th Friend by Marjorie Weinman Sharmat

The Three Little Pigs (Traditional)

What Comes in 2s, 3s, and 4s? by Susan Aker

WHAT TIME IS IT?

A family spends the hours of the day getting ready for a special visitor.

Level 1

THE GROUCHY LADYBUG
Time story role play

A bright idea from Kimberly Jordano, kindergartner teacher, Lee School, Los Alamitos, California

After reading *The Grouchy Ladybug*, have children work in groups to create headbands for each character in the story. Group members can wear their headbands as they act out what happens each hour of the day, and audience members can show the hour of the day on individual clocks. Challenge children to work in groups to create their own version of a very grouchy creature who learns a lesson by the end of the day.

Materials
- ✓ *The Grouchy Ladybug* by Eric Carle
- ✓ construction paper
- ✓ individual clocks
- ✓ art supplies

MATH CONCEPTS
- connecting units of time to real-life events
- experiencing durations of time
- measuring: *time*
- recognizing numerals
- reading analog clock faces
- sequencing by time

WRITING FRAMES
What time is it?

It's ____ o'clock.

It's time to ____.

At ____ o'clock, I ____.

At *five* o'clock, I *watch my favorite cartoon*.

MAGNET CLOCK
Hands-on clock

Cut clock hands from magnetic strips and place them in a center with magnetic numbers and a metal pizza pan. Include *What Time Is It?* and other books about time, such as *Bear Child's Book of Hours*. As children read the books, they can create the different times with clock manipulatives.

Materials
- ✓ magnetic strips
- ✓ magnetic numbers
- ✓ metal pizza pan
- ✓ *What Time Is It?*
- ✓ *Bear Child's Book of Hours* by Anne Rockwell

RELATED SKILLS
- punctuality
- contractions: *it's, o'clock*
- punctuation: *question marks, periods*
- question and answer format
- sequencing

OUR SCHOOL SCHEDULE
Daily schedules

Discuss the school day schedule. Record each activity and its starting and ending time on two sentence strips. Include a simple illustration or photo depicting the activity. Place one set of sentence strips in a pocket chart in correct sequence. Select a time keeper for each time period and give each the extra copy of the sentence strip to keep on his or her desk. Time keepers are responsible for keeping the class on schedule. When time is up for the scheduled activity, time keepers ring the bell, announce the time, and the next activity. At the end of the week, ask children what they learned about schedules, and have them record their ideas in their journals. Assign new time keepers each week. Make extra sentence strips for special events such as assemblies, library time, or computer lab. As an extension activity, children can practice sequencing the sentence strips in the pocket chart.

Materials
- sentence strips
- pocket chart
- clock stamp and ink pad
- small bell
- journals
- art supplies

REBUS CLOCK SYSTEM
Daily activities

Help each child make a clock constructed as follows:

1. Draw one circle 4" in from the edge of the construction paper circle.

2. Draw little lines on the inner circle to mark and label the hour positions.

3. Attach "hands" to the center of each clock with a brad.

Have children keep their clocks with them and draw pictures on their clocks of what they do at different times during the day. They can begin at 12 noon (at school) and record events until 6:00 pm (at home). They can start recording again at 7:00 am the next day until noon. When children share their clocks with the class, they can move the clock hands to match the activity they are describing.

Materials
- 12" construction paper circles
- 6" and 8" construction paper clock hands
- brads
- art supplies

MY FAVORITE TIME OF DAY
Class book

A bright idea and project from Jennifer Botenhagen and her first graders, Lu Sutton School, Novato, California

Ask children to share their favorite time of day and why they like it. Have them write a sentence about this time, explaining their choice. Invite children to illustrate and use the rubber clock stamp to show the time on their page. Compile individual pages into a class book of time for children to read during free time. Place a clock with moveable hands in a pocket in the book so as children read each page, they can match the time.

Materials
- construction paper
- drawing paper
- clock stamp and ink pad
- clock with moveable hands
- art supplies

40 Learn to Read Resource Guide • Math Creative Teaching Press

LEARNING A SKILL

Punctuality

Read the book *Guy Who Was Five Minutes Late*, a humorous, rhyming story about a baby who was born five minutes late and grows up being five minutes behind for everything. In their journals, have children write about or draw activities that require punctuality. Discuss why it is important to be on time. Divide children into groups of four or five and give each group chart paper labeled *On Time* and *Late*. Have children write the positive consequences of being on time and those of being late. Ask each group to put together a skit with simple props about one of the things they wrote about. Invite each group to share their lists and act out their skits for the class.

Materials
- ✓ *Guy Who Was Five Minutes Late* by Bill Grossman
- ✓ journals
- ✓ chart paper
- ✓ art supplies

LINKING SCHOOL TO HOME

Family book

Send home a blank book and directions sheet with each child, asking his or her family to create their own book modeled after *What Time Is It?* Suggest family members draw or take pictures of each other getting ready for a special event or spending a weekend day together. Have them record the time each picture was taken. Families can use the writing frame *It's _____ o'clock. It's time to _____*.

Materials
- ✓ blank books
- ✓ directions sheets
- ✓ camera (optional)

LITERATURE LINKS

All in a Day by Mitsumasa Anno

At This Very Minute by Kathleen Rice Bowers

Bear Child's Book of Hours by Anne Rockwell

Clocks and How They Go by Gail Gibbons

Clocks and More Clocks by Pat Hutchins

The Completed Hickory Dickory Dock by Jim Aylesworth

The Grouchy Ladybug by Eric Carle

Guy Who Was Five Minutes Late by Bill Grossman

How Do You Say It Today, Jesse Bear? by Nancy White Carlstrom

Nine O'Clock Lullaby by Marilyn Singer

Once Upon a Time by Gwenda Turner

A Seed, a Flower, a Minute, an Hour by Joan Blos

Time for Bed by Mem Fox

Time to . . . by Bruce McMillan

Level I

WHO TOOK THE COOKIES FROM THE COOKIE JAR?

Delightful pictures illustrate this fun adaptation of the traditional clapping chant.

MATH CONCEPTS

- story problems
- pattern counting by twos
- subtraction

WRITING FRAMES

"Who took the ____ from the ____?"

"____ took the ____ from the ____."

"Who took the <u>ice cream</u> from the <u>ice cream cone</u>?"

"<u>Shana</u> took the <u>ice cream</u> from the <u>ice cream cone</u>."

RELATED SKILLS

- high-frequency words: *who, the, from*
- question and answer format
- punctuation: *quotation marks*
- speech bubbles

TWO BY TWO
Dramatizing with puppets or masks
Counting by twos

A bright idea and project from Nancy Hoyt and her second graders, Abraham Lincoln School, Long Beach, California

Materials
- ✓ paper plates
- ✓ cotton balls
- ✓ pipe cleaners
- ✓ construction paper scraps
- ✓ craft sticks
- ✓ art supplies

After practicing counting by twos and clapping the pattern of the story, assign children to cooperative groups. Have each group create one mask or puppet from the story—queen, cow, snake, sheep, giraffe, cat, and mouse. They can use cotton balls to make fluffy sheep, pipe cleaners for cat and mouse whiskers, and long construction paper giraffe necks. Have each child also make and decorate two construction paper cookies. Invite children to act out the story with their masks or puppets. Place "cookies" in a large cookie jar, and have the "mouse" sneak out two cookies at a time and give them to the "snake." Invite children to write sentences about the cookies with *two* as the main number.

Learn to Read Resource Guide • *Math* — Creative Teaching Press

ACTIVITY

WHO TOOK THE COOKIES?
Story dramatization

Provide each child with a cookie-jar-shaped work mat made from construction paper, two napkins, and ten small cookies. Read *Who Took the Cookies from the Cookie Jar?*, and have children remove cookies two at a time, hiding them between their napkins. As you read the last page, substitute children's names for the snake, for they will soon be eating the cookies as a special snack!

As an alternative idea, place ten seasonal candies or treats in a fish bowl and have children chant the story, using the treats in place of the cookies. For example, use candy corn in the fall and conversation hearts in February.

As an extension, model writing equations from the book and have children write equations on individual chalkboards. In a learning center, place a copy of the book, ten blocks to represent cookies, and a sock to represent the snake. As children read the story, the "snake" can gobble up the "cookies."

Materials
- small cookies or cookie cereal
- construction paper
- napkins
- *Who Took the Cookies from the Cookie Jar?*
- individual chalkboards
- blocks
- sock

ACTIVITY

HOW MANY FEET IN THE BED?
Number problems
Counting by twos

Read *How Many Feet in the Bed?* and as a follow-up, ask children how many feet are in all the beds in their houses when everyone is in bed. Have children make simple folded houses using 12" x 18" construction paper and decorate the outsides of their houses. They can then open them up and draw all the beds and feet in their own houses. Invite children to practice counting by twos when counting all the feet.

Materials
- *How Many Feet in the Bed?* by Diane Johnston Hamm
- 12" x 18" construction paper
- art supplies

Creative Teaching Press · Learn to Read Resource Guide • *Math*

LEARNING A SKILL

Story problems

Model story problems as children act them out with real cookies. For example: *Silvia peeked in the cookie jar and saw eight cookies. She took three out and ate them. How many were left in the jar?* Show children how to write the equation on a chalkboard.

Materials
- small cookies
- cookie-jar-shaped work mats
- individual chalkboards
- drawing paper
- art supplies

Invite children to generate stories for their classmates to solve, using real cookies. Model how to record equations, and encourage children to write the appropriate equations on their chalkboards. They can work with partners to solve more difficult story problems. For example: *Monday morning there were eight cookies, but Monday night there were only four cookies left. How many cookies were eaten on Monday?* After children work on problems with cookies or counters, have them share and explain their strategies.

Children can use art supplies to illustrate their own story problems, writing equations on the back. Assemble stories in a class book and keep it in a center with salt-dough cookies for independent problem solving.

LINKING SCHOOL TO HOME

Counting by twos
Retelling the story

Place the book, baby food jar, ten pennies or buttons, and sock in the plastic bag. Children can take the bag home to retell the story with their parents, using the objects in the jar to represent cookies being eaten by the "snake" (sock).

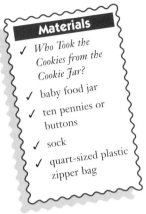

Materials
- *Who Took the Cookies from the Cookie Jar?*
- baby food jar
- ten pennies or buttons
- sock
- quart-sized plastic zipper bag

LITERATURE LINKS

A Bag Full of Pups by Dick Gackenbach

The Doorbell Rang by Pat Hutchins

A House for Hermit Crab by Eric Carle

How Many Feet in the Bed? by Diane Johnston Hamm

How Many Twos? by Judy Hindley

The Napping House by Audrey Wood

Odds and Evens by Heidi Goennel

The Shopping Basket by John Burningham

Two Ways to Count to 10 by Ruby Dee

What Comes in 2s, 3s, and 4s? by Suzanne Aker

Level II

A-COUNTING WE WILL GO

Children count colorful animals in this fun counting song and rhyme.

PATTERN COUNTING
Rhyming words
Class counting book

A bright idea and project from Kathy Dahlin and her first graders, Cooper School, Superior, Wisconsin

Brainstorm rhyming words for numbers such as *one/sun, two/shoe, three/key, four/door, five/hive,* and *ten/pen*. Use these rhyming words to create an adaptation of *A-Counting We Will Go*. For example:

A-counting we will go, a-counting we will go.
We'll count the suns
And group them by ones,
And then we'll count each row.

Write new text on sentence strips and place them in a pocket chart for children to read. Children can illustrate each strip with drawings, stickers, and magazine pictures. Compile pages into a class book for children to practice pattern counting.

Materials
- sentence strips
- pocket chart
- construction paper
- stickers
- magazines
- art supplies

MATH CONCEPTS

- comparing quantities
- counting
- sight recognition of sets up to five
- matching number words to quantities
- number concepts
- one-to-one correspondence

WRITING FRAMES

We'll count the ____
And ____ them ____
____,
And then we'll let them go.
We'll count the ants
And put them on plants,
And then we'll let them go.

RELATED SKILLS

- phonics: *rhyming words; ou sound (count)*
- contraction: *we'll*
- plural forms: *-s (bugs), -es (foxes)*

Creative Teaching Press — Learn to Read Resource Guide • Math

ONE-TO-ONE LEARNING CENTER
Concrete math
One-to-one correspondence

Write selected verses from *A-Counting We Will Go* on tagboard cards. Gather small manipulatives to match objects in the song, such as:

- plastic bugs with "rugs" (small fabric pieces)
- penny-prize frogs and "logs" (twigs)
- cat counters with felt mats
- craft bees and felt trees
- small bears and doll house chairs

Materials
- ✓ 7" x 11" tagboard cards
- ✓ manipulatives
- ✓ shoe box
- ✓ art supplies

Place manipulatives and cards in a decorated box in a learning center. Children can choose a card to read or sing and then find the matching manipulatives to pair up and count.

A-COUNTING HEADBANDS
Character headbands
Circle singing game

Sing *A-Counting We Will Go* to the tune of *A-Hunting We Will Go*. Help children make sentence strip headbands with construction paper pictures of the rhyming words in the song. Each child can make one headband and a matching prop. Invite children to sit in a circle and wear their headbands as they sing and reenact the story.

Materials
- ✓ sentence strips
- ✓ construction paper
- ✓ art supplies

NUMBER FLIPS
Number stories

Provide each child with a sentence strip to complete as follows: *How do you make ____?* At the end of each strip, staple five or six squares. Let children illustrate number combinations with stickers, cutout magazine pictures, and small objects. Staple children's sentence strips together to make a tactile class book of number combinations.

Materials
- ✓ sentence strips
- ✓ 3" construction paper squares
- ✓ stickers
- ✓ magazines
- ✓ small objects (popcorn, beans, buttons, gravel)
- ✓ art supplies

LEARNING A SKILL

Rhyming words

Have children work in groups of four or five to make rhyming word wheels to go with the story. Children may draw pictures or glue on pictures from magazines. Have them write the name of each story character and several rhyming words. Invite children to take turns spinning the spinner. If the spinner lands on *goats*, the child reads all the rhyming words and chooses one pair to sing. For example: *A-counting we will go, a-counting we will go. We'll count the goats and put them in coats, and then we'll let them go!*

Materials
- ✓ 12" tagboard circles
- ✓ spinners
- ✓ brads
- ✓ magazines
- ✓ art supplies

LINKING SCHOOL TO HOME

Take-home bag

Decorate a backpack or canvas bag with fabric markers and puffy paint. Place a copy of *A-Counting We Will Go* in the bag with several manipulatives from the *One-to-One Learning Center* activity. In a parent letter, describe the math concepts children are exploring. Invite families to read and sing the book together and practice one-to-one matching with manipulatives in the bag. Also, ask family members to find or draw one set of objects to count and match and place them in the paper bag for the children to share at school. Suggestions include: toy bears and chairs, noodles (spiral macaroni) and poodles (drawings), or cats and hats.

Materials
- ✓ *A-Counting We Will Go*
- ✓ backpack or canvas bag
- ✓ fabric markers
- ✓ puffy paint
- ✓ manipulatives
- ✓ parent letter
- ✓ small paper bags

LITERATURE LINKS

Anno's Counting Book by Mitsumasa Anno

Counting on Frank by Rod Clement

Counting Wildflowers by Bruce McMillan

Each Orange Had 8 Slices by Paul Giganti, Jr.

From One to One Hundred by Teri Sloat

One Cow Moo Moo by David Bennett

Ten Black Dots by Donald Crews

Who Wants One? by Mary Serfozo

Who's Counting? by Nancy Tafuri

Level II

THE BUGS GO MARCHING

Adorable bugs march across the pages in this fun adaptation of the traditional counting song.

MATH CONCEPTS

- sequencing numbers
- attributes
- connecting words to quantities
- multiplication
- pattern counting
- sorting and classifying

WRITING FRAMES

The ____ go marching ____ by ____,

A ____ one stopped to ____.

And they all go marching, ____ they go!

The <u>kids</u> go marching <u>two</u> by <u>two</u>,

A <u>sneezy</u> one stopped to go, "<u>Achew!</u>"

And they all go marching, <u>through the school</u> they go!

RELATED SKILLS

- parts of speech: *action words (verbs), descriptive words (adjectives)*
- phonics: *rhyming words*

NUMBER LADYBUGS
Number combinations

A bright idea and project from Sally Griffith and her kindergartners and first graders, Carillo School, Westminster, California

Help children make ladybugs with moveable wings:

1. Trace and cut out ladybug wings from red construction paper.
2. Trace and cut out the ladybug body from black construction paper.
3. Attach wings to the body with brads.
4. Attach construction paper "legs" to each side of the black body and two antennae to the head. Curl antennae around a pencil. Use paper scraps to make eyes.

Assign a number being studied (such as five) and have children use beans or blocks to make several combinations of that number. Ask each child to choose one combination to make on his or her ladybug. Have them attach black circles to the wings to represent the number combination. Children can open the wings and write their chosen number with white crayon on the black body under the wings. Use ladybugs to practice addition and show various addition facts.

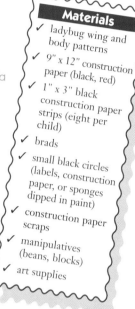

Materials
- ✓ ladybug wing and body patterns
- ✓ 9" x 12" construction paper (black, red)
- ✓ 1" x 3" black construction paper strips (eight per child)
- ✓ brads
- ✓ small black circles (labels, construction paper, or sponges dipped in paint)
- ✓ construction paper scraps
- ✓ manipulatives (beans, blocks)
- ✓ art supplies

48 Learn to Read Resource Guide • Math Creative Teaching Press

BEAN BUGS
Hands-on counting

A bright idea from Sally Griffith, K–1 teacher, Carillo School, Westminster, California

Materials
- ✓ large bag of lima beans
- ✓ spray paint (red, yellow)
- ✓ fine-line permanent markers
- ✓ newspaper

Place lima beans in a single layer on newspaper and spray yellow paint on one side. When dry, turn beans over and spray the other side red. When dry, children can draw little black dots from one to ten on the red side of the beans. They can arrange beans in twos, fours, sixes, and so on to match and count the bugs on each page in *The Bugs Go Marching*. Use the bean bugs as manipulatives for addition, subtraction, word problems, and skip-counting activities.

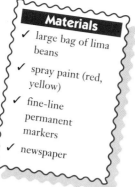

BUGS, BUGS, BUGS
Interactive counting book

Have children refer to *The Bugs Go Marching* as they illustrate a similar background on tagboard pieces to create a free-standing, accordion-fold big book. Tape tagboard pieces together and add text. Have each child make and paint two or three clay or salt-dough bugs. As children read each page, have them place the correct number of bugs in front of the background. Children can also draw bugs and staple them to popsicle sticks to use as bug pointers as they read or sing the rhyme.

Materials
- ✓ 11" x 17" tagboard pieces
- ✓ plastic tape
- ✓ clay or salt dough (1 part salt, 2 parts flour, 1 part water)
- ✓ drawing paper
- ✓ popsicle sticks
- ✓ art supplies

BUG MASKS
Story retelling and counting with bug masks

A bright idea and project from Cathy Young and her kindergartners and first graders, Biella School, Santa Rosa, California

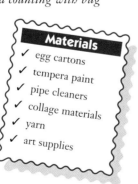

Materials
- ✓ egg cartons
- ✓ tempera paint
- ✓ pipe cleaners
- ✓ collage materials
- ✓ yarn
- ✓ art supplies

After children are familiar with the song from *The Bugs Go Marching*, create bug masks using painted egg carton eyes, pipe cleaner antennae, and colorful collage material decorations. Use yarn to tie masks on children's heads. Have two "bugs" stand in front of the class. As the class begins singing the song, each "bug" picks one more "bug" to sing the *four by four* part. Those children then pick two more "bugs" to sing the *six by six* part, and so on. Play the game again and again!

LEARNING A SKILL

Sequencing numbers

Give groups of two or three children a sentence strip with text from one page of the story, a green construction paper "hill," and a bag of bugs. Ask each group to read their sentence strip and move into position around the room in the order of the book. Have them lay sentence strips on the floor with the correct number of bugs. When all groups finish, invite them to "story walk" from group to group, singing and reading the story while looking at each display.

Materials
- ✓ sentence strips from *The Bugs Go Marching*
- ✓ green construction paper
- ✓ plastic or paper bugs
- ✓ 10 plastic zipper bags

LINKING SCHOOL TO HOME

Bug story problems

Have children take turns bringing home the blank book, plastic bugs, and *The Bugs Go Marching*. They can use the plastic bugs to act out the story with their families, then draw bugs on their assigned page and decorate the background. Display the book in the classroom, and invite children to share their contributions.

Materials
- ✓ *The Bugs Go Marching*
- ✓ large blank book with text only
- ✓ plastic bugs
- ✓ art supplies

LITERATURE LINKS

The Amazing Anthony Ant by Lorna and Graham Philpot

How Many Feet in the Bed? by Diane Johnston Hamm

One Hundred Hungry Ants by Elinor J. Pinczes

1 Hunter by Pat Hutchins

Six Creepy Sheep by Judith Ross Enderle and Stephanie Gordon Tessler

Six Sleepy Sheep by Jeffie Ross Gordon

Six Snowy Sheep by Judith Ross Enderle and Stephanie Gordon Tessler

The 329th Friend by Marjorie Weinman Sharmat

Two by Two by Barbara Reid

2 x 2 = Boo! A Set of Spooky Multiplication Stories by Loreen Leedy

Level II

THE CRAYOLA® COUNTING BOOK

Colorful crayons inspire children to count by ones, twos, fives, and tens.

 ACTIVITY

THE COLORS OF OUR WORLD
Class color book

A bright idea and project from Tebby Corcoran and her kindergartners, Lee School, Los Alamitos, California

Materials
- ✓ 12" x 18" construction paper
- ✓ magazines
- ✓ stickers
- ✓ art supplies

Cut crayon-shaped pages from 12" x 18" sheets of white, red, yellow, orange, green, blue, purple, brown, and black construction paper. Have children find pictures from magazines, color their own pictures, or find stickers to attach to each corresponding book page. Add photos of children and the environment to each page, along with short descriptive sentences.

As an extension, add a pocket to the left-hand side of each two-page spread of the class color book. When children read the book during independent reading, ask them to write a story problem using objects shown on the right page. Invite them to write the answer on the back, and place the story problem in the pocket. The next person to read the book can solve the problem and write a new one for the next reader to solve.

MATH CONCEPTS
- place value
- comparing size
- counting by ones, twos, fives, and tens
- patterns
- sorting by color
- tallying

WRITING FRAMES

Count the ____ crayons.

Count by ____.

Count the crayons that ____.

Count the <u>orange</u> crayons.

Count by <u>fours</u>.

Count the crayons that <u>have pointy tops</u>.

RELATED SKILLS
- **phonics:** *rhyming words, vowel digraphs (ou, ow), consonant blends (scr, gr, br)*
- **vocabulary:** *color words, size words*

Creative Teaching Press — Learn to Read Resource Guide • *Math*

COUNTING CRAYONS
Learning center activity

Materials
- ✓ The Crayola® Counting Book
- ✓ crayons
- ✓ shoe box
- ✓ plastic zipper bags

Place a copy of *The Crayola® Counting Book* at a math center with a shoe box full of crayons. Children working at the center can use crayons from the box to make arrangements shown on pages 4, 5, 8, 10, and 11 of the book. Have them count the crayons in their arrangements. Challenge children to develop strategies for counting every crayon in the book. They can write word problems using crayons and place the problems with the crayons needed to solve them in plastic bags for others to solve.

CRAYON MEASURING
Measurement comparisons

Materials
- ✓ new crayons
- ✓ recording sheets
- ✓ art supplies

Give each pair of children a box of crayons and a recording sheet. Children can use the crayons to measure objects in the classroom and record their findings on a recording sheet. After measuring at least five objects, have children draw and write about objects using the frame *A _____ is as long as _____ crayons.* Add pictures and text to a class poster or graph.

GRAB AND GRAPH IT!
Making a "real" graph

A bright idea from Sally Griffith, K-1 teacher, Carrillo School, Westminster, California

Materials
- ✓ new boxes of eight crayons
- ✓ masking tape

Invite each child to choose a favorite color from his or her box of crayons and hide it in a pocket or sock. Ask children to sit in a circle on the floor. Place a strip of masking tape on the floor to make the bottom of a "graph." Arrange all the crayons from a box under the tape, allowing 5" between each crayon. Have individual children place their chosen crayon above the masking tape line. Tally chosen colors on the chalkboard. When the floor graph is complete, discuss the results, including information you can and cannot get from the graph.

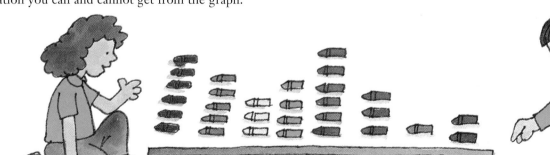

LEARNING A SKILL

Place value

Invite children to bring crayons from home to donate to the class. Label the 5-gallon bucket *hundreds*, the 1-gallon bucket *tens*, and the beach bucket *ones*. Place them on a table at the front of the room with a pile of rubber bands. As each child comes up to contribute crayons, the class counts each crayon dropped into the *ones* bucket and keeps a tally on the erasable board. When there are ten crayons, the child bundles them and places them in the *tens* bucket. Erase the *ones* tally and start a new tally in the *tens* column. When the tally reaches one hundred, bundle the crayons and place them in the *hundreds* bucket, then erase the *tens* tallies and make a *hundreds* tally. Continue until all crayons are counted, bundled, and recorded. Write the final total on the board and let children see all the crayons in the buckets. Relate this activity to known stories about big numbers such as *When the Doorbell Rings* and *The 329th Friend*.

Materials
- crayons
- 5-gallon bucket
- 1-gallon bucket
- beach bucket
- rubber bands
- erasable board
- *When the Doorbell Rings* by Pat Hutchins
- *The 329th Friend* by Marjorie Weinman Sharmat

LINKING SCHOOL TO HOME

Colors at home

Give each child a sheet of construction paper. Help children fold their papers to form eight squares and label each square with a color. At home, children can keep a tally of the colors they find, then total their tallies when finished. On the back of their papers, invite children to explain their totals. For example, families may have favorite colors or decorations for holidays or the season.

Materials
- 12" x 18" construction paper
- crayons (yellow, red, orange, purple, blue, brown, black, green)

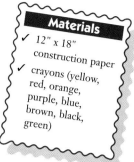

LITERATURE LINKS

Colors Everywhere by Tana Hoban

How Many Snails? by Paul Giganti, Jr.

Is It Red? Is It Yellow? Is It Blue? by Tana Hoban

The M&M's Chocolate Candies Counting Book by Barbara Barbieri McGrath

Mouse Count by Ellen Stoll Walsh

Mouse Paint by Ellen Stoll Walsh

Of Colors and Things by Tana Hoban

Pete's Chicken by Harriet Ziefert

Planting a Rainbow by Lois Ehlert

Rainbow Round-A-Bout by Ernest Nister

Splash! by Ann Jonas

The 329th Friend by Marjorie Weinman Sharmat

12 Ways to Get to 11 by Eve Merriam

When the Doorbell Rings by Pat Hutchins

Level II

FIVE LITTLE MONSTERS

Silly monsters help bring to life this traditional counting poem.

MATH CONCEPTS

- connecting quantities to number words up to five
- counting backward and forward
- number concepts
- number patterns
- sequencing
- subtraction

WRITING FRAMES

_____ little _____ swinging from the tree,
_____ called _____ and _____ agreed . . .

Five little <u>children</u> swinging from the tree,
<u>Auntie</u> called <u>Uncle</u> and <u>Uncle</u> agreed . . .

RELATED SKILLS

- comparing and contrasting
- parts of speech: *action words (verbs), descriptive words (adjectives)*
- phonics: *rhyming words*
- punctuation: *quotation marks, commas, exclamation points, periods*

APPLE PICKING TIME
Subtraction

A bright idea and project from Kathy Dahlin and her first graders, Cooper School, Superior, Wisconsin

Materials
- flannelboard apple tree with apples, numbers, and symbols
- sentence strips
- pocket chart
- art supplies

Adapt *Five Little Monsters* to create a flannelboard story entitled *Five Little Apples*. Have children take turns removing flannelboard apples as the class says each verse and makes the corresponding number sentences. Provide each child with two sheets of paper. On one page, have children draw or trace a tree and cut five slits for apples; on the other page, have children draw or trace five apples with tags to cut out and place in the tree. Children can take away one apple at a time as the class repeats each verse of the apple story.

Learn to Read Resource Guide • Math

Creative Teaching Press

ACTIVITY: LOTS OF LITTLE MONSTERS
Interactive story board

A bright idea and project from Renee Keeler and her second graders, Lee School, Los Alamitos, California

Materials
- ✓ construction paper
- ✓ magnetic board or flannelboard
- ✓ small magnets or Velcro strips
- ✓ large mug
- ✓ art supplies
- ✓ collage materials

Have children work in pairs to make small monsters to use as characters for retelling the story. Laminate and mount monsters on magnets or Velcro strips. Create a storyboard background by having children paint and cut out pictures to attach to a magnetic board or flannelboard. Write number sentences and word problems from *Five Little Monsters* on construction paper cards. Invite children to take turns drawing cards from a mug and using their monsters to tell and solve the number sentences and word problems.

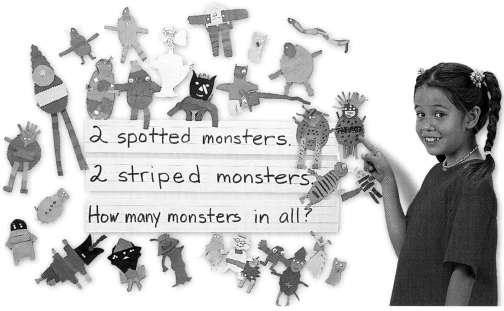

ACTIVITY: FIVE LITTLE ANIMALS
Class big book adaptation

Brainstorm ideas for text adaptations using different animals. For example:

*Five little cows running in the corn,
One fell down and bumped his horn.*

*Six little dinosaurs swinging from the tree,
One fell off and bumped his knee.
T-Rex called Apatasaurus and Apatasaurus agreed,
"No more dinosaurs swinging from the tree!"*

*Five little spiders hanging from a tree,
One crawled away—how many do you see?*

Materials
- ✓ chart paper
- ✓ construction paper
- ✓ art supplies

Invite children to share their ideas for rhyming words as you write the new text on chart paper. Print copies of the new text on sheets of construction paper and have children work in groups to decorate each page with colorful animals. Assemble pages in a class big book for sharing and counting.

Creative Teaching Press   Learn to Read Resource Guide • Math   55

LEARNING A SKILL

Comparing and contrasting

Gather several versions of the book, such as *Five Ugly Monsters* by Tedd Arnold or *Five Little Monkeys Jumping on the Bed* by Eileen Christelow. Read class-created versions as well. After reading several versions, choose two versions to compare and contrast on a Venn diagram.

As an extension, assign children to cooperative groups, and ask each group to choose a version of the book to dramatize with simple props for the class.

Materials
- ✓ several versions of *Five Little Monsters*
- ✓ chart paper

LINKING SCHOOL TO HOME

Crafts
Counting

Send home materials in paper bags with each child. With help from family members, invite children to create five funny monster stick puppets. They can decorate their monsters with collage materials such as tissue paper, yarn, and cotton balls. Ask parents to help children place their monsters in different locations around the house to demonstrate positions such as *under the table*, *in the kitchen*, and *on the sofa*. Send home the text from *Five Little Monsters* so children can read it to their parents and demonstrate the subtraction math in the story. Children can bring their puppets to school to use for puppet shows.

Materials
- ✓ paper bags
- ✓ construction paper
- ✓ craft sticks
- ✓ art supplies
- ✓ collage materials
- ✓ text from *Five Little Monsters*

LITERATURE LINKS

Counting Sheep by John Archambault

Don't Forget the Bacon! by Pat Hutchins

Five Little Ducks by Ian Beck

Five Little Monkeys Jumping on the Bed by Eileen Christelow

Five Little Monkeys Sitting in a Tree by Eileen Christelow

Five Ugly Monsters by Tedd Arnold

One Bear with Bees in His Hair by Jakki Wood

1 Hunter by Pat Hutchins

One Yellow Lion by Matthew Van Fleet

Seven Little Hippos by Mike Thaler

Where the Wild Things Are by Maurice Sendak

Let's Measure It!

Level II

A tabby cat who loves to measure fish invites children to measure the lengths of everything!

FISHY LENGTHS
Measuring lengths

A bright idea and project from Anne Seapy and her first graders, Los Alamitos Elementary, Los Alamitos, California

Read and discuss *Let's Measure It!* and related books such as *Blue Sea* and *Fish Eyes: A Book You Can Count On*. Have each child create an underwater background scene, using a wide brush or sponge to wash over drawing paper with water and quickly painting the surface with mixtures of blue and green watercolors. When dry, have children cut out several fish of various lengths from construction paper or wallpaper scraps. Children can use 1" tiles or a ruler to create at least one fish of each length from 1" to 6".

Have children paste their fish onto their watercolor backgrounds, adding other construction paper details such as seaweed, snails, and shells. Attach finished artwork to large sheets of butcher paper, leaving space at the bottom. Children can measure and cut out their own drawings or magazine pictures and attach them to the butcher paper under the questions *What else is ____ inches long?* Title the display *Let's Measure It!*, and place standard and nonstandard units of measurement near the display so children can measure and compare fish and other objects.

Materials
- ✓ *Let's Measure It!*
- ✓ *Blue Sea* by Robert Kalan
- ✓ *Fish Eyes: A Book You Can Count On* by Lois Ehlert
- ✓ watercolor paints
- ✓ wide brushes or sponges
- ✓ drawing paper
- ✓ construction paper or wallpaper scraps
- ✓ 1" tiles or rulers
- ✓ blue butcher paper
- ✓ art supplies

MATH CONCEPTS
- linear measuring: *standard units*
- comparing sizes

WRITING FRAMES

The ____ in the ____ is ____ inch/es long.

What else is ____ inch/es?

The <u>cat</u> in the <u>pet shop</u> is <u>eight</u> inches long.

What else is <u>eight</u> inches?

RELATED SKILLS
- asking questions
- contraction: *let's*
- high-frequency words: *what, the, is*
- plural form: *-es (inches)*
- punctuation: *question marks*

Creative Teaching Press — Learn to Read Resource Guide • *Math*

HANDS-ON MEASUREMENT
Measuring lengths

A bright idea and project from Debbie Martinez and her kindergartners, Helen Lehman School, Santa Rosa, California

Have children trace and cut out their hand shapes from construction paper. Using the hand shapes, invite children to measure objects around the classroom and say to a partner, *It takes (number) hands to measure a (object).* Display hands on the bulletin board.

Materials
- construction paper squares (size of a child's hand)
- art supplies

MEASURING UP
Measuring and graphing

Provide children with 6" tagboard rulers they can hole-punch, string with yarn, and wear around their necks.

Assign children inch measurements from one to six and have them search the classroom for several objects measuring the assigned lengths. Children can share, graph, and discuss their findings with the class.

Materials
- tagboard
- yarn
- hole punch

LOTS TO MEASURE!
Learning center activity

Make a chart with two columns entitled *Let's Measure It!* Label the left column from one to six inches.

Photocopy pages 3, 5, 7, 9, 11, 13, 14, and 15 from the book. Children can color the pictures, cut them out, measure them, and glue them to the appropriate chart sections. Children can also measure and cut out pictures from toy catalogs to add to the chart.

Materials
- *Let's Measure It!* chart
- toy catalogs
- 6" rulers
- art supplies

LEARNING A SKILL

Measuring and sequencing lengths

Invite children to bring a favorite stuffed animal or doll to class. (Provide extras if needed.) Sit in a circle on the floor and brainstorm similarities between stuffed animals, such as length, color, and clothing. Help children make a floor graph by sorting toys by length and labeling each group as *short*, *medium*, and *long*. Then have children collect their toys one category at a time. Have children use 1" cubes or rulers to determine the lengths of their toys. Have them write lengths on masking tape to stick on their toys. Invite children to work together to sequence their toys from shortest to longest.

Materials
- ✓ stuffed animals
- ✓ 1" cubes or rulers
- ✓ masking tape

LINKING SCHOOL TO HOME

Measuring
Graphing and sorting lengths

Encourage parents to provide their children with rulers and share ways to estimate an inch. Ask parents to help children find 1", 2", and 3" items to bring to school. Remind them not to send anything valuable, dangerous, or breakable. At school, have children place their items on the graph labeled *1"*, *2"*, and *3"*. Compare and discuss the lengths of the items, and ask how children found their lengths. Place items in a tub with the labeled graph for children to sort and measure on their own.

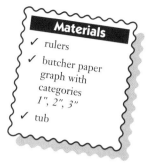

Materials
- ✓ rulers
- ✓ butcher paper graph with categories 1", 2", 3"
- ✓ tub

LITERATURE LINKS

A Big Fish Story by Joanne and David Wylie

The Biggest Nose by Kathy Caple

Blue Sea by Robert Kalan

Farmer Mack Measures His Pig by Tony Johnston

Fish Eyes: A Book You Can Count On by Lois Ehlert

How a House Is Built by Gail Gibbons

How Big Is a Foot? by Rolf Miller

How Big Were the Dinosaurs? by Bernard Most

Inch by Inch by Leo Lionni

Is the Blue Whale the Biggest Thing There Is? by Robert E. Wells

Jim and the Beanstalk by Raymond Briggs

The Littlest Dinosaurs by Bernard Most

May I Bring a Friend? by Beatrice Schenk de Regniers

Ocean Parade by Patricia MacCarthy

Slower Than a Snail by Anne Schreiber

The Snake: A Very Long Story by Bernard Waber

Level II

LITTLE NUMBER STORIES: ADDITION

Colorful photographs illustrate number combinations up to ten.

MATH CONCEPTS

- counting by ones, twos, and tens
- logical thinking: *exploring strategies for solving story problems*
- number concepts: *addition (combining two or more sets)*
- using numerical equations to describe combining two or more sets

WRITING FRAMES

If you have ____ ____ and ____ ____, what do you have?

You have ten ____!

If you have three spiders, two worms, and five snails, what do you have?

You have ten creepy crawlies!

RELATED SKILLS

- high-frequency words: *if, you, have, what*
- parts of speech: *descriptive words (adjectives)*
- question and answer format
- number words
- sorting and classifying

 FRUIT DAY
*Classifying
Graphing*

A bright idea and project from Linda Benton and her first graders, Westwood School, Napa, California

Materials
- variety of fruit
- chart paper
- drawing paper
- graphing plastic or butcher paper grid
- art supplies

Have children sit in a circle and brainstorm various ways to sort the fruit. Choose one idea and have children group the fruit accordingly. For example, they might sort by color or how the fruit grows (trees, bushes). Discuss how it can be difficult to tell which group has more fruit without actually counting. Ask how they can arrange the fruit so it is easier to see which group has more without counting (put them in rows). Introduce the graphing plastic (or butcher paper grid) and sort the fruit into straight rows. Observe how much easier it is to compare groups without actually counting. Model how to combine and find the difference between rows of fruit on the graph. For example: *How many apples and oranges do we have altogether? How many fewer pears are there than apples?* Invite children to write and draw number stories about the fruit in their math journals. As a follow-up, children can make fruit prints to illustrate their number stories. Assemble these "fruitful" equations in a class book or on a bulletin board.

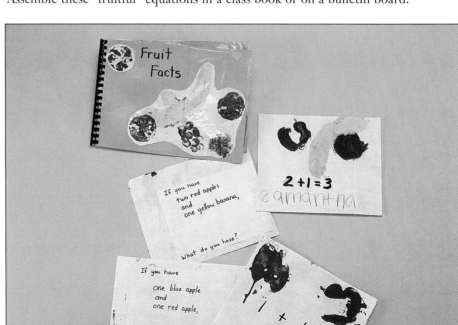

LET'S PARTY!
Class book

The next time you are planning a class party, celebration, or event, photograph the preparation and organization of the supplies. Photograph the supplies so they can be easily counted. Use the photographs for a class book. Add text and have children write word problems about the objects in the photos for others to solve.

Materials
- ✓ camera
- ✓ blank book
- ✓ art supplies

If you have 22 cups,
22 napkins,
22 cupcakes,
lots of M&Ms,
21 children,
and 1 teacher,
what do you have?
A class party!

THE BIG PIZZA BOX BOOK
Addition equations

A bright idea and project from Majella Maas and her kindergartners, Abraham Lincoln School, Long Beach, California

Paint the inside and outside of the pizza box or cover it with construction paper. Cut the tagboard so it fits inside the box. Attach tagboard "pages" to the pizza box with brads. Create story problems for your class to illustrate, using art supplies, construction paper, and paint. Place manipulatives in the pizza box book for children to use when reading the book on their own.

Materials
- ✓ large pizza box
- ✓ tagboard
- ✓ brads
- ✓ neon paint
- ✓ construction paper
- ✓ art supplies

LITTLE "IF" STORIES
Creative writing
Describing words
Classifying

Assign children to collaborative groups of four or five. Give each group the last line of an *If You Have . . .* story. Here are some suggestions:

You have 15 happy farm animals!
You have 12 cool birthday presents!
You have 11 scary monsters!
You have 10 creepy crawlies!

Materials
- ✓ writing paper
- ✓ art supplies
- ✓ collage materials

Each group's task is to write and illustrate the story that leads up to their assigned last lines. For example:

If You Have . . .
If you have 2 cute caterpillars,
3 slimy snails,
4 sparkling spiders,
and 1 goofy grasshopper,
what do you have?
You have 10 creepy crawlies!

Group members can illustrate their stories and combine the pages into a book to share with the class. Place all the books in a box for all to enjoy after the presentations.

LEARNING A SKILL

Counting by tens
Estimating

Before starting this activity, send home a letter asking parents to help their child place ten small objects in a plastic bag and bring them to school. Suggest objects such as plastic bugs, toy dinosaurs, cars, shells, or rocks. Children must not bring anything breakable, valuable, or dangerous. Have children sit in a circle on the floor, holding their own bag of objects. Ask, *How many objects do you think there are altogether?* Record estimates on a chart and ask for the lowest and highest estimates. Write these numbers at the bottom of the chart with a line between them and explain that this is the "range" of estimates. Ask, *How can we find out for sure if the actual number falls in this range? Do we need to count everything in every bag? Why or why not?* Have each child check to see if there are exactly ten items in his or her bag.

Children can also trade and count items in a friend's bag. As the class counts by tens around the circle, have each child hold up his or her bag or deposit it into a basket or tub (ten bags per tub). Write the numerals on the chart in order (10, 20, 30, 40 . . .), and ask children if they notice a pattern. Check to see if the resulting number falls within the class's range of estimates. Discuss what children learned and how they would estimate differently next time.

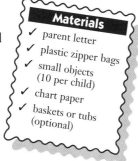

Materials
- ✓ parent letter
- ✓ plastic zipper bags
- ✓ small objects (10 per child)
- ✓ chart paper
- ✓ baskets or tubs (optional)

LINKING SCHOOL TO HOME

Story problems

Ask parents to help children use their toys to create several story problems and explain them to a family member. Have children choose one problem to illustrate and write about. Children can bring their stories to school for sharing and explain how they solved their problems.

Materials
- ✓ toys
- ✓ writing paper
- ✓ art supplies

LITERATURE LINKS

Adding Animals by Colin Hawkins

Anno's Counting Book by Mitsumasa Anno

Anno's Counting House by Mitsumasa Anno

A Dozen Dogs: A Read and Count Story by Harriet Ziefert

Eating Fractions by Bruce McMillan

Happy Birthday, Sam by Pat Hutchins

A House for Hermit Crab by Eric Carle

The M&M's Brand Chocolate Candies Counting Book by Barbara Barbieri McGrath

So Many Cats! by Beatrice Schenk de Regniers

Something Absolutely Enormous by Margaret Wild

10 Black Dots by Donald Crews

12 Ways to Get to 11 by Eve Merriam

Little Number Stories: Subtraction

Level II

A variety of common objects are grouped together by color and function.

MY OWN LITTLE NUMBER STORIES
Subtraction paper bag book

A bright idea and project from Marcia Fries and her kindergartners, Lee School, Los Alamitos, California

Materials
- ✓ small paper bags
- ✓ construction paper
- ✓ small objects (cereal, buttons, dry beans, plastic bugs)
- ✓ binding material (yarn, rings, staples)
- ✓ art supplies

Gather small objects and help children write a number story to go with each set. Type stories and give each child a copy. Children can paste stories on a paper bag "page" and make illustrations using construction paper patterns such as shirts and cereal bowls. Have them also attach construction paper answer flaps. Bind each child's bags together to make a book. Children can read and act out the stories, using the corresponding small objects. Store objects in bags for future use.

MATH CONCEPTS

- using numerical equations to describe separating sets
- counting forward and backward
- logical thinking: *exploring strategies for solving story problems*
- number concepts: *subtraction (separating sets)*

WRITING FRAMES

If you had ten ____ and ____, what happened to ten?

There are ____ ____ left.

If you had ten pizzas *and* four were eaten, *what happened to ten?*

There are six pizzas *left.*

RELATED SKILLS

- cause and effect
- classifying
- high-frequency words: *if, you, had, what, there, are*
- question and answer format
- number words

READY! SET! ACTION!
Class photograph book

Divide children into small groups and have each group design their own subtraction number story using the sentence frame from the text. For example: *If there were 12 children and 4 went home with earaches, what happened to 12? There are 8 children left.* Photograph children acting out their stories. Help them use magnetic numbers on the backs of cookie sheets to make number sentences that complement their stories. Photograph number sentences for story solutions so children can see themselves in action!

Materials
- ✓ camera
- ✓ art supplies
- ✓ magnetic numbers
- ✓ cookie sheets

MORE NUMBER FUN
Addition and subtraction learning center

Place manipulatives and a copy of the book in a learning center. Children at the center can manipulate the props as they read the book. They can also write their own addition and subtraction problems using manipulatives or pictures in the book.

Materials
- ✓ *Little Number Stories: Subtraction*
- ✓ 10 empty juice boxes (assorted flavors)
- ✓ 10 balloons
- ✓ 10 sandwich pictures
- ✓ 10 plastic bananas
- ✓ writing paper

LIKE MAGIC!
Subtraction slider cards

Make slider cards for each child as follows:

1. Fold 6" x 12" construction paper in half lengthwise.
2. Glue edges across the top and down one side to make a pocket.
3. Cut a 2 1/2" x 12" construction paper strip to fit inside.

Give children stickers and have them place half on the pocket and the other half on the slider strip. Children can make up number stories and slide the card in to make pictures disappear to provide a visual representation of subtraction.

Materials
- ✓ 6" x 12" construction paper
- ✓ 2½" x 12" construction paper strips
- ✓ stickers
- ✓ art supplies

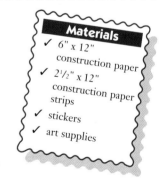

LEARNING A SKILL

Using numerical equations to describe separating sets

Gather small objects and make a story board to go with each set of objects. Ideas might include pumpkin seeds on a pumpkin, buttons on a shirt, ants on a picnic table of food, and rings on a hand. Write subtraction number sentences on index cards, and place all the cards and objects for each story board in a small container. Children can build and tell story problems to go with number sentences.

Materials
- ✓ small objects (plastic ants, spiders, rings, dry beans)
- ✓ story boards
- ✓ index cards
- ✓ small storage containers

LINKING SCHOOL TO HOME

Number stories

Display picnic basket items and have children create number stories to go with them. Record stories in a picnic-basket-shaped class book, and have children illustrate the stories. Leave blank pages in the book so children can record new stories for items in the basket. Place inside the picnic basket: picnic items, magnetic letters, a small cookie sheet, picnic-basket-shaped class book, and *Little Number Stories: Subtraction*. Send the basket home with a different child each night. Children can spread the tablecloth on a table or floor and act out the stories with items in the basket. Have each child record his or her stories using the magnetic letters on the cookie sheet and add new stories to the class book. Note: Include a directions sheet in the class book for parents.

Materials
- ✓ picnic basket
- ✓ picnic basket items (tablecloth; paper plates; napkins; plastic forks, spoons, and knives; cups; plastic food; empty yogurt containers)
- ✓ magnetic letters
- ✓ small cookie sheet
- ✓ picnic-basket-shaped class book
- ✓ *Little Number Stories: Subtraction*
- ✓ directions sheet

LITERATURE LINKS

A Bag Full of Pups by Dick Gackenbach

A House for Hermit Crab by Eric Carle

How Many Snails? by Paul Giganti, Jr.

Math in the Tub by Sara Atherlay

Rooster's Off to See the World by Eric Carle

The Shopping Basket by John Burningham

Splash! by Ann Jonas

Take Away Monsters by Colin Hawkins

The Tiny Seed by Eric Carle

What's Left? by Judi Barrett

Level II

Lunch with Cat and Dog

Cat and Dog explore fractions as they share their food.

MATH CONCEPTS

- equal shares
- conservation of volume
- equivalent fractions

WRITING FRAMES

"____! I want the most!" said Cat.

"OK," said Dog. "____ for you and ____ for me."

"*A chocolate bar! I want the most!*" said Cat.

"OK," said Dog. "*Four pieces for you and one piece for me.*"

RELATED SKILLS

- learning manners
- comparing: *big/little, most/least*
- punctuation: *quotation marks*
- thought and speech bubbles
- writing conversation

 CAT AND DOG'S PIZZA BOX FRACTIONS
Hands-on class book
Exploring fractions

A bright idea and project from Candice Siu and her first graders, Lee School, Los Alamitos, California

Help children create an interactive book as follows:
1. Use construction paper scraps to decorate the outside of the pizza box and five tagboard circles to represent five different kinds of pizzas.
2. Cut five file folders to approximately 8" x 10½". Tape both sides of the folders closed to form pockets.
3. Staple all five pockets together on the left and glue the bottom one to the inside lid of the pizza box.
4. Label folders with the days of the week and the different kinds of pizzas children made.
5. Glue the remaining tagboard circle to the inside bottom of the pizza box.
6. Divide children's "pizzas" into fractional pieces. For example, cut one in half, thirds, fourths, and so on. Laminate pizza pieces and place them in the appropriate pockets.
7. Make a 3" x 8" book entitled *Who Got the Most?* Book pages should correspond with the pizzas in the pockets and contain text following this frame:

On <u>Monday</u>, Cat and Dog had <u>pineapple</u> pizza. Dog had <u>three</u> pieces and Cat had <u>one</u> piece. Who got the most?

Glue the little book next to the tagboard circle in the bottom of the pizza box.

Invite children to match pizza pieces from one pocket with a page in the little book, manipulating the fractional pieces to answer the questions. Children can make up their own story problems and place them in the pockets with the pizza pieces.

Materials
- pizza box
- construction paper scraps
- 5 file folders
- six 8" tagboard circles
- art supplies

ACTIVITY: CAT WILL BE CAT!
Creative writing
Learning manners

Discuss good manners and being polite. Compare Cat's behavior in the three *Learn to Read* books. Refer to Jack Gantos' *Rotten Ralph* series and compare Cat to Ralph to discover how they are the same and different. Read *It's a Spoon, Not a Shovel!*, a very enjoyable book about etiquette, showing comical animals in situations requiring good manners. Invite children to create an etiquette book for Cat entitled *Cat's Book of Good Manners*.

Materials
- ✓ Lunch with Cat and Dog
- ✓ All Through the Week with Cat and Dog by Rozanne Lanczak Williams
- ✓ Cat and Dog by Rozanne Lanczak Williams
- ✓ Rotten Ralph series by Jack Gantos
- ✓ It's a Spoon, Not a Shovel! by Caralyn Buehner
- ✓ blank book
- ✓ art supplies

ACTIVITY: FRACTION MOBILES
Edible pizza fractions

A bright idea and project from Eileen Young and her first and second graders, Abraham Lincoln School, Long Beach, California

Have children work with a partner to bake their own pizzas with English muffins, spaghetti sauce, and cheese. Challenge children to divide their pizzas using different fractions (number of pieces).

Have children work with their partners to trace, cut out, and decorate a table shape on construction paper. Help children make a flap from the bottom half of the tablecloth (see photograph) and write a statement under the flap, such as *"Pizza! I want the most!" yelled Cecilia*. Children can then illustrate the pieces of pizza they ate and hang them from the table bottom with string. On the back of each piece, have children write how much they have. For example: *Javier has one piece and Cecilia has four pieces*.

Materials
- ✓ English muffins
- ✓ spaghetti sauce
- ✓ cheese
- ✓ toaster oven or microwave
- ✓ plastic knives
- ✓ table-shaped patterns
- ✓ 8½" x 11" construction paper
- ✓ hole punch
- ✓ string
- ✓ art supplies

LEARNING A SKILL

Equal shares
Learning manners

Divide the class into groups of four and give each group the suggested materials. Instruct group members to work together to divide the food into four equal shares and then eat their fair share. Ask one person from each group to tell the rest of the class about how they shared the food.

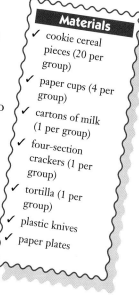

Materials
- cookie cereal pieces (20 per group)
- paper cups (4 per group)
- cartons of milk (1 per group)
- four-section crackers (1 per group)
- tortilla (1 per group)
- plastic knives
- paper plates

LINKING SCHOOL TO HOME

Family fractions

Provide each child with paper and an envelope. Ask parents to have their children observe as they cut a pizza, pie, cake, or bread into fractions for the whole family. Ask them to point out how they fill glasses half full or how they divide items equally. They can write and draw about how their family divides food into equal shares. Invite children to bring their ideas and pictures to school in the envelope to share with the class.

Materials
- drawing paper
- envelopes

LITERATURE LINKS

Eating Fractions by Bruce McMillan

Fraction Action by Loreen Leedy

Gator Pie by Louise Mathews

A Giraffe and a Half by Shel Silverstein

Half-birthday Party by Charlotte Pomerantz

It's a Spoon, Not a Shovel! by Caralyn Buehner

The Little Mouse, the Red, Ripe Strawberry, and the Big Hungry Bear by Audrey Wood

Pezzettino by Leo Lionni

Picture Pie: A Circle Drawing Book by Ed Emberley

Rotten Ralph series by Jack Gantos

Sharing by Taro Gomi

THE MAGIC MONEY BOX

Magicians use a "magic" box to experiment with money equivalencies.

 THE MAGIC MONEY BOX BIG BOOK
Interactive big book

A bright idea and project from Marcia Fries and her kindergartners, Lee School, Los Alamitos, California

Materials
- ✓ 12" x 18" construction paper
- ✓ money (real, pictures, or stamps)
- ✓ laminated pockets
- ✓ plastic zipper bag
- ✓ art supplies

Cut construction paper into box-shaped pages. Have children write the text for each page using the frame *In go(es) ____. Out come(s) ____.* Use the chart from *The Magic Money Box* activity to help make equal money combinations. Ask a small group of children to decorate a construction paper peek-over figure and glue it to the back of the book. Have other groups of children decorate the money box pages using construction paper, glitter, paint, and other art supplies.

Staple a plastic bag on the title page to hold the "money." Paste laminated pockets on the book's pages. As children read the book, they can take money from the plastic bag, match the correct amount of coins with the text, and place them in the laminated pockets.

Level II

MATH CONCEPTS
- money values: *equivalencies and counting*

WRITING FRAMES

In go ____ ____ and ____ ____.

Out come/s ____ ____!

In go three nickels and one dime.

Out comes one quarter!

RELATED SKILLS
- opposites: *in/out*
- punctuation: *exclamation points*
- money words

Creative Teaching Press   Learn to Read Resource Guide • *Math*

ACTIVITY: THE MAGIC MONEY BOX

Class money box
Coin equivalencies

Materials
- shoe box
- small box lid
- colored self-adhesive paper
- Velcro strips
- coins
- magic wand
- chart paper
- laminated pockets
- art supplies

Create a Magic Money Box as follows:

1. Cover the lid and bottom of the shoe box with self-adhesive paper.
2. Cut a slit in the lid.
3. Cover the smaller lid with self-adhesive paper and attach it to the inside of the shoe box lid with Velcro.
4. Each day, prepare for the Magic Money Box activity by choosing equal coin amounts such as five pennies and one nickel. Place one of the "amounts" (nickel) inside the box, and have the other "amounts" (five pennies) readily available.

When "magic money time" arrives, have a child put five pennies into the box. Print the amount put in the money box on the left side of the chart (see illustration), and let children predict what coin will be inside the box. Wave a magic wand and say some "magic" words. Carefully remove the lid without showing its underside and reveal the nickel in the box. Place the nickel in a laminated pocket on the chart and record the amount. Remove the money in the "secret lid" to use for future magic shows.

ACTIVITY: MINI MAGIC MONEY BOOKS

Individual money books

A bright idea and project from Karen Joseph and her first graders, Lee School, Los Alamitos, California

Materials
- camera
- magician dress-up clothes
- 6" x 18" construction paper strips
- 2" construction paper squares
- sticker stars
- glitter glue
- coins
- art supplies

Take a picture of each child dressed as a magician in front of the class money box. Provide each child with two strips of construction paper. Help children fold each strip to make four accordion-style pages and tape these together to form eight pages. Demonstrate how to trim off the top and bottom corners to form a box shape.

Show children how to draw lines to complete a box shape. Fold the edges of 2" squares and paste the folds to the books' pages to make lift-up flaps. Children can use the frame *In go(es) ____. Out come(s) ____* to write their own text, making equal combinations. They can use coin stamps or real coins to illustrate their texts. To complete Magic Money Books, have children glue their photographs to the covers and decorate them with stars and glitter glue.

LEARNING A SKILL

Money values—equivalencies

Provide children with sentence strips and have them cut pictures from catalogs. To make puzzles, have each child paste a picture at one end of his or her sentence strip. Children can then attach a white square with a price to the picture. Invite children to use coin stamps to make equal money amounts on the other side of their pictures. Have children cut the sentence strip apart in zig-zag fashion and place pieces in a tissue box. They can decorate their tissue boxes with construction paper, stickers, glitter, colored glue, and markers. Invite children to share their puzzle banks with classmates, matching money amounts and fitting puzzle pieces.

Materials
- sentence strips
- catalogs
- 1" white squares
- coin stamps
- tissue boxes
- construction paper
- art supplies

LINKING SCHOOL TO HOME

Magic with money

Place the Magic Money Box, magician's hat, wand, and copy of *The Magic Money Box* in a backpack for children to take home and share with their families. Children can use real coins to demonstrate the Magic Money Box to family members. Have them see if their families can guess the magic!

Materials
- Magic Money Box (see page 70)
- magician's hat
- wand
- *The Magic Money Box*
- backpack
- coins

LITERATURE LINKS

Alexander, Who Used to Be Rich Last Sunday by Judith Viorst

Arthur's Funny Money by Lillian Hoban

Arthur's Honey Bear by Lillian Hoban

A Chair for My Mother by Vera B. Williams

Department Store by Gail Gibbons

The Go-around Dollar by Barbara Johnston Adams

How the Second Grade Got $8,205.50 to Visit the Statue of Liberty by Nathan Zimelman

The Money Book and Bank by Elaine Wyatt and Stan Hinden

One Magic Box: A Book about Numbers by Roger Chouinard and Mariko Chouinard

Pigs Will Be Pigs by A. Axelrod

A Quarter from the Tooth Fairy by Caren Holtzman

"Smart" from *Where the Sidewalk Ends* by Shel Silverstein

Something Special for Me by Vera B. Williams

26 Letters and 99 Cents by Tana Hoban

Whirr Pop Click Clang by Carol Mackenzie

Level II

SPIDERS, SPIDERS EVERYWHERE!

Lovable spiders appear in different places for readers to count.

MATH CONCEPTS

- counting
- logical thinking
- one-to-one correspondence
- problem solving
- sorting and classifying

WRITING FRAMES

There are ____ spiders ____.

There are ____ ____ ____.

There are <u>nine spiders by the ice cream cones</u>.

There are <u>eight</u> <u>puppies on the bed</u>.

RELATED SKILLS

- parts of speech: *prepositional phrases*
- high-frequency words: *there, are*
- phonics: *rhyming words, consonant blend sp*
- vocabulary: *position words*

HERE COME THE SPIDERS!
Interactive story boards

A bright idea and project from Renee Keeler and her second graders, Lee School, Los Alamitos, California

Write the sentences from *Spiders, Spiders Everywhere!* on individual sentence strips, leaving a blank where the number appears. Write the numerals or words *one* through *ten* on word cards that fit in the sentence blanks. Assign children to groups of two or three, and give each group a sentence strip and tagboard piece. Brainstorm ways groups can create backgrounds to illustrate their sentences using art supplies and collage materials. Suggestions may include: using yarn for a spider web and hair, fabric for a bedspread and curtains, tissue paper for fire in a fireplace and tree leaves, construction paper for a door that opens with a button doorknob, and clear plastic for a window.

When background boards are finished, place them in a learning center with the plastic spiders, sentence strips, number word cards, and a copy of *Spiders, Spiders Everywhere!* At the learning center, have children sequence story boards on the floor, place sentence strips with corresponding boards, choose number words for each sentence, and place the right number of spiders in each scene.

Materials
- ✓ sentence strips
- ✓ twelve 11" x 14" blue tagboard pieces
- ✓ word cards
- ✓ 55 plastic or child-made spiders
- ✓ *Spiders, Spiders Everywhere!*
- ✓ art supplies
- ✓ collage materials

Learn to Read Resource Guide • Math

SILLY SPIDER STORIES
Hands-on math

A bright idea and project from Gina Lems-Tardif and her first graders, Lee School, Los Alamitos, California

Before starting the activity, create a spider web on the black felt with white fabric paint. Cut 2" Styrofoam balls in half and spray-paint the 1" and 2" Styrofoam balls. Help children attach Velcro to the flat bottom of each 2" half-ball. They can use glue and toothpick halves to attach spider heads. Help children add eight spider legs by sticking pipe cleaners into each Styrofoam body. Invite them to glue on wiggly eyes and decorate their spiders. Challenge children to manipulate the spiders in the felt web to demonstrate addition and subtraction word problems or other spider songs and chants.

Materials
- square yard of black felt
- white fabric paint
- spray paint
- 1" Styrofoam balls
- 2" Styrofoam balls
- sticky Velcro squares
- toothpicks
- pipe cleaners (cut into 2" pieces)
- wiggly eyes
- art supplies

THUMBPRINT SPIDERS
Spider counting books and grids

A bright idea and project from Liz Newman and her first graders, Hidden Valley School, Santa Rosa, California

Show children how to make thumbprint spiders. After a little practice, children can make the correct number of thumbprint spiders on each page for a "one through ten" counting book. On the last page, children can use mini spider stickers to complete a grid like that on page 16 of *Spiders, Spiders Everywhere!*

Materials
- blank student books
- ink pads (blue, red, green)
- spider stickers
- art supplies

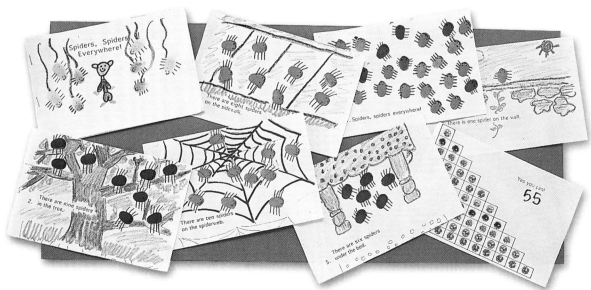

LEARNING A SKILL

Prepositional phrases

Write prepositional phrases from the book on sentence strips and place them in random order in the pocket chart. As you read *Spiders, Spiders Everywhere!*, stop after each prepositional phrase and have a volunteer find it in the pocket chart. When finished, all the phrases should be in order.

Materials
- sentence strips
- pocket chart
- *Spiders, Spiders Everywhere!* big book
- blank books
- art supplies

Give each child a blank book and invite him or her to write and illustrate a rhyming book featuring a favorite book character, stuffed animal, or pet. Children can refer to the pocket chart for an ending phrase for each page.

LINKING SCHOOL TO HOME

Spiders class book

Have each child gather materials to make a spider and place them in a paper bag. Ask parents to help children use materials to create a spider. They can also use craft materials from home. Invite children to bring their spiders to school in their paper bags and share how they worked with family members. Take photographs of the spiders in various places around the classroom, such as behind the door, under a desk, in the wastebasket, and on the window ledge. Glue photographs into a book and print words on each page, using the language pattern from the text. Add the title *Spiders Invade Room ___*. Encourage children to borrow the book overnight to share with family members.

Materials
- yarn
- pipe cleaners
- egg cartons
- Styrofoam balls
- buttons
- dry pasta
- dry beans
- small paper bags
- camera

LITERATURE LINKS

Be Nice to Spiders by Margaret B. Graham

Count! by Denise Fleming

Counting Wildflowers by Bruce McMillan

Deep Down Underground by Olivier Dunrea

The Icky Bug Counting Book by Jerry Pallotta

The Itsy Bitsy Spider by Iza Trapani

One Cow Moo Moo by David Bennett

One, Five, Many by Kveta Pacovska

Pigs from 1 to 10 by Arthur Geisert

Ten Black Dots by Donald Crews

Ten Cats Have Hats by Jean Marzollo

Ten Little Rabbits by Virginia Grossman

The Very Busy Spider by Eric Carle

Ten Monsters in a Bed

The number of funny monsters in bed keeps decreasing in this innovative adaptation of the traditional poem.

Level II

MONSTERS, MONSTERS, MONSTERS
Story problem role play

A bright idea and project from Gina Lems-Tardif and her first graders, Lee School, Los Alamitos, California

Have children work in groups of two or three to design a headband representing each monster in the story as well as the spider. They can cut the character faces from fluorescent-colored tagboard and add features using puffy paint and construction paper scraps. Help children attach monster faces to tagboard headbands. Write the subtraction equations on sentence strips.

Invite eleven children at a time to take turns wearing the headbands and acting out the story while the audience reads with you. The ten "monsters" can stand holding the blanket as one monster at a time "rolls off" the "bed" and falls on some floor pillows. The "spider's" role is to change the equation card to correspond with each story problem. Each actor can choose another child until all have participated.

Materials
- fluorescent-colored tagboard
- puffy paint
- construction paper scraps
- tagboard strips
- sentence strips
- art supplies
- bed sheet or blanket
- equation cards

MATH CONCEPTS
- matching numerals with words
- connecting quantities to number words up to ten
- counting backward and forward
- number concepts
- subtraction

WRITING FRAMES

There were ____ in a bed and the ____ one said, "Roll over, roll over!"

They all rolled over and one fell out.

There were <u>seven</u> in a bed and the <u>jolly</u> one said, "Roll over, roll over!"

They all rolled over and one fell out.

RELATED SKILLS
- high-frequency words: *there, were, in, they, said*
- parts of speech: *descriptive words (adjectives)*
- punctuation: *quotation marks, exclamation points*

MONSTER SUBTRACTION
Subtraction with paper monsters

A bright idea and project from Marcia Smith and her kindergartners and first graders, Lu Sutton School, Novato, California

Materials
- ✓ 12" x 18" drawing paper
- ✓ 8½" x 11" drawing paper
- ✓ art supplies

Distribute one large and one small sheet of drawing paper to each child. Help children fold the larger papers into ten equal parts. After reading *Ten Monsters in a Bed*, have children draw, color, and cut out one monster from each section for a total of ten. Children may want to match monsters in the book, reinforcing describing words by making one mad, blue, sad, funny, spotted, round, shy, green, pink, and striped. Have children draw, color, and cut out bed shapes from their smaller sheets of paper. As you reread the story, invite children to follow along, using their monsters and bed shapes.

MOVE OVER MONSTERS
Addition wall story

Discuss whether the text shows adding or subtracting monsters (subtracting).

Materials
- ✓ chart paper
- ✓ butcher paper
- ✓ art supplies

Have children create an addition monster story, following the predictable text. Brainstorm ten describing words for the monsters in the wall story and create the text. For example: *There was/were (number) in a bed and the (adjective) one said, "Move over, move over!" They all moved over and one jumped in.* Pre-cut butcher paper into bed-shaped pages for children to illustrate with monsters.

TEN ON THE BUS
Pocket chart activity

A bright idea and project from Esta Peterson and her second and third graders, Hoover School, Yakima, Washington

Materials
- ✓ tagboard
- ✓ class pictures
- ✓ sentence strips
- ✓ word cards
- ✓ art supplies

Cut tagboard to fit in a pocket chart and use it to make a bus with cutout windows to accommodate students' pictures. Tape an extra piece of tagboard to the back of the bus for a sliding card. Make one or more sliding cards with students' pictures. Using sentence strips and word cards, have children help create an innovation of *Ten Monsters in a Bed* entitled *Ten Children on a Bus*. After practicing the new song with the class several times, place the pocket chart in a center for children to read and sing on their own.

LEARNING A SKILL

Matching numerals with words

Cut two circles from tagboard and divide both into ten pie sections. Print number words *one* through *ten* on one of the circles and have children place monster stickers, one to ten, in the sections of the other circle. Cut a bed shape from tagboard. Cut two windows in the bed the size of the pie sections on the circles. Have children decorate the bed by gluing on fabric squares. Attach the circles to the windows with brads. Reproduce the text in the book and attach it to the tagboard bed.

Invite children to retell the story, turning the wheels so the number of monsters matches the number words.

Materials
- ✓ tagboard
- ✓ brads
- ✓ monster stickers
- ✓ small fabric squares
- ✓ art supplies

LINKING SCHOOL TO HOME

Story retelling

Send directions sheets home, asking parents to help children create and decorate beds using tissue boxes and collage materials. Cut out the bottom of the boxes and have children decorate five white paper squares as five monsters from the story. Wrap each monster picture around a finger and tape it in place. Ask children to place their hands under the "bed" and hold up five fingers as they retell the story. Family members can also make equation cards for each "scene." Encourage parents and children to change roles. Invite children to bring their project to school to perform for the class.

Materials
- ✓ directions sheets
- ✓ rectangular tissue boxes
- ✓ 2" white paper squares
- ✓ construction paper
- ✓ art supplies
- ✓ collage materials

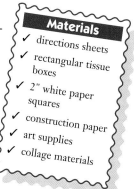

LITERATURE LINKS

Five Little Monkeys Jumping on the Bed by Eileen Christelow

Five Ugly Monsters by Tedd Arnold

Nine Ducks Nine by Sarah Hayes

One Hungry Monster by Susan Heyboer O'Keefe

One Was Johnny by Maurice Sendak

The Right Number of Elephants by Jeff Sheppard

Seven Little Hippos by Mike Thaler

Seven Little Rabbits by John Becker

Six Sleepy Sheep by Jeffie Ross Gordon

Ten Apples up on Top by Theo LeSieg

Ten Little Mice by Joyce Dunbar

Ten Little Rabbits by Virginia Grossman

Ten Sly Piranhas by William Wise

Too Many Monkeys by Kelly Oechsli

When One Cat Woke Up: A Cat Counting Book by Judy Astley

Level II

THE TIME SONG

Learn units of time measurement with this catchy time song!

MATH CONCEPTS

- measuring: *units of time*

WRITING FRAMES

____ ____ in one ____.

Time for/to ____.

Twenty-four hours in one day.

Time to run and read and play.

RELATED SKILLS

- days of the week
- phonics: *rhyming words*
- seasons of the year
- sorting and classifying

Activity

OUR CLOCK BOOK
Class book

A bright and idea and project from Kimberly Jordano and her kindergartners, Lee School, Los Alamitos, California

Gather a variety of watches, clocks, and toy clocks and display them at an exploration center. Discuss how the clocks are the same or different and compare them with clocks in *What Time Is It?* Have children use individual clocks as you (or volunteers) read the story.

On large sheets of construction paper, have children draw a picture of their favorite activity and record what time of day it occurs. Collect pages and assemble them into a class book. Make a giant clock with moveable hands for the cover. Take photographs of children during the school day and place them in a class book with appropriate sentences for children to share. For example: *At 8:00 am Kim comes to school* or *At 8:30 am Tony listens to a story.*

Materials
- variety of clocks
- individual clocks
- *What Time Is It?*
- construction paper
- brads
- art supplies
- camera

Learn to Read Resource Guide • Math   Creative Teaching Press

IT'S SHOW TIME!
Story dramatization

Help children create, practice, and perform a simple musical show for another class, upper-grade buddies, or parents. Divide children into seven groups. One group is the chorus and the rest perform six acts (the other verses of the song). The chorus sings the first two lines of the song before each act and hold up props to symbolize the highlighted time in that act. For example, the chorus can introduce Act II, pages 6–7 of the book, by singing:

Sing this song and sing this rhyme.
Learn some ways we measure time!

As they sing, have children hold up a wristwatch, alarm clock, and digital clock.

The actors assigned to Act II sing:

Sixty minutes in one hour.
Time to wake up, eat, and shower.

Have children pantomime waking up, eating breakfast, and taking a shower.

Materials
- ✓ *The Time Song* big book
- ✓ props
- ✓ time-measuring items (watches, egg timers, clocks, calendars)

TIME TO . . .
Time mural

A bright idea and project from Kimri Vella and her first graders, Lee School, Los Alamitos, California

Materials
- ✓ construction paper
- ✓ brads
- ✓ magazines
- ✓ sentence strips
- ✓ art supplies

Draw a large clock face and add moveable hands. Display the clock on a bulletin board. Add moveable clock hands. Invite children to search magazines for pictures that represent events and attach times corresponding to the pictures. For example, attach a card showing 6:00 pm to a picture of a family eating dinner. Help children write a phrase on a sentence strip to match each picture, such as *It's time to eat dinner*. Invite each child, one at a time, to choose a picture, read the time and phrase, and arrange the clock hands to match.

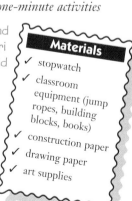

IN ONE MINUTE!
Pop-up book of one-minute activities

A bright idea and project from Karri Haven and her second graders, Helen Lehman School, Santa Rosa, California

Materials
- ✓ stopwatch
- ✓ classroom equipment (jump ropes, building blocks, books)
- ✓ construction paper
- ✓ drawing paper
- ✓ art supplies

Brainstorm things children can do in one minute, then ask each child to pick something he or she would like to do. Have children do their activities for one minute. After discussing the activities, ask each child to draw a small picture about what he or she did. Children can also write or dictate sentences about their pictures. Use the illustrations and writing to create a class pop-up book entitled *Things We Can Do In One Minute*.

Creative Teaching Press • Learn to Read Resource Guide • Math

LEARNING A SKILL

Understanding a time line

Attach a length of shelf paper to the chalkboard and add the title *One Day in Room ____*. Divide the time line into half-hour sections. As you progress through the day, stop every half hour and have children help write the current activity on the time line. The next day, assign children to small groups and have each group refer to the time line to complete a page for a big book about the class schedule.

Materials
- ✓ blank shelf paper
- ✓ blank class book
- ✓ art supplies

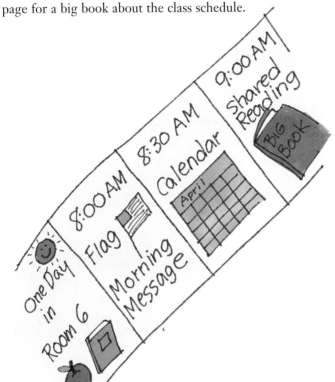

LINKING SCHOOL TO HOME

Memorable moments in time

A bright idea and project from Peggy Fretz and her second graders, Mulberry School, Whittier, California

Materials
- ✓ 12" x 18" construction paper
- ✓ art supplies

Give each child three sheets of construction paper. Help children tape their paper together to make a picture story entitled *Memorable Moments in the Life of (student's name)*. Ask parents to help their children recall special times and illustrate them in sequence, showing dates, photographs (if available), illustrations, and special details. Encourage children to bring the picture stories to school for sharing.

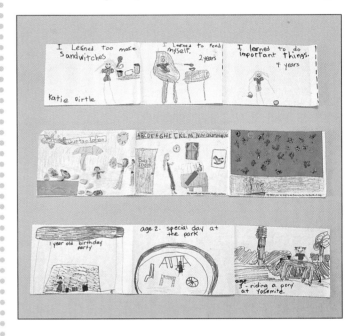

LITERATURE LINKS

Busy Monday Morning by Janina Domanska

Chicken Soup with Rice by Maurice Sendak

Clocks and More Clocks by Pat Hutchins

How Do You Say It Today, Jesse Bear? by Nancy White Carlstrom

Nine O'Clock Lullaby by Marilyn Singer

Red Day, Green Day by Edith Kunhardt

The Seasons of Arnold's Apple Tree by Gail Gibbons

A Seed, a Flower, a Minute, an Hour by Joan Blos

Time for Bed by Mem Fox

Time to . . . by Bruce McMillan

Turtle Day by Douglas Florian

Wednesday Is Spaghetti Day by Maryann Cocca-Leffler

When This Box Is Full by Patricia Lillie

While I Sleep by Mary Calhoun

The Wonderful Counting Clock by Cooper Edens

A Year in the Country by Douglas Florian